FACING THE INTERNATIONAL ENERGY PROBLEM, 1980-2000

In cooperation with

The Center for
Strategic and International Studies
Georgetown University
Washington, D.C.

FACING THE INTERNATIONAL ENERGY PROBLEM, 1980-2000

Amos A. Jordan

Hayden Bryan

Michael Moodie

PRAEGER

PRAEGER SPECIAL STUDIES • PRAEGER SCIENTIFIC

Published in 1979 by Praeger Publishers
A Division of Holt, Rinehart and Winston/CBS, Inc.
383 Madison Avenue, New York, New York 10017 U.S.A.

© 1979, the Center for
Strategic and International Studies

9 038 987654321

Library of Congress Catalog Number: 79-54706

Printed in the United States of America

PREFACE

More than 60 years ago, Frederick Soddy observed:

No-one today is ignorant of the part played by
energy, not only in science, but in industry,
politics and the whole science of human welfare.
From the cradle to the grave everyone is dependent
on Nature for an absolutely continuous supply of
energy in one or other of its numerous forms.
When the supplies are ample there is prosperity,
expansion, and development. When they are not,
there is want. (*Matter and Energy*, 1912.)

Supplies of petroleum are highly uncertain; nuclear
technology is extraordinarily controversial; coal is environ-
mentally suspect. In the fact of such difficulties, will
there be sufficient energy supplies or not? Power balances,
the solidarity of the Western Alliance, the development
plans of newly independent nations, the vitality of the
international economic system all hinge on the answer to
this central question of energy adequacy.

The Fourth Quadrangular Conference, "Facing the
International Energy Problem, 1980-2000," co-chaired by
Senator William V. Roth, Jr., and Congressman Al Ullman,
Chairman of Ways and Means, was held at the Center for
Strategic and International Studies from May 8 to 10, 1978.
Bringing together international experts from government,
academia, and business, the conference examined both the
range of energy issues confronting the international system
and the spectrum of policy options available to societies
to meet those issues. The resulting findings were so
responsibly reasoned and carefully expressed that the Center
wishes to bring them through this publication to a wider
audience.

As Henry Kissinger observed at the inaugural session
of the CSIS Future of Business Program in 1977, the most
decisive global issues "are at the intersection of economics
and politics, of national and international policies, of
the public and private sectors. It is essential that both
policymakers and businessmen understand these important
new interrelationships."

The topic of energy is clearly a decisive global issue such as Secretary Kissinger described. Accordingly, the comprehensive analyses contained in the following report are recommended reading for policymakers, businessmen, and concerned citizens in general. The events of late 1978 and early 1979, most notably the upheaval in Iran, have not invalidated the analysis in the various chapters of this volume, which was written before they occurred. Rather, the upheaval in oil markets and the roughly 60 percent increase in oil prices, which resulted from supply inadequacies arising out of the Iranian situation, have made more urgent many of the lessons underscored here.

Amos A. Jordan
Executive Director

TABLE OF CONTENTS

INTRODUCTION

The Arab oil embargo in 1973 and the ensuing series of
steep increases in oil prices by OPEC began to make the
American public aware of the vital role energy, particularly
oil, plays in the U.S. economy. Energy shortages, escalating
inflation, and a mounting trade deficit caused increasing
concern and raised difficult questions.

In the last five years, no one has really provided the
simple answers about energy--because there are none. That
much Congress has learned.

The outlook for the next quarter of a century's energy
is uncertain.

Fixing the number of barrels of oil or cubic feet of
gas beneath the surface of this planet is an inexact science
at best. The man on the street finds it miraculous that
President Carter's speech on the moral equivalent of war is
so quickly followed with spectacular reports of new oil
discoveries.

Mexico now claims proven reserves of at least 20 billion
barrels. China hints at massive untapped fields. The North
Sea is just rushing on stream. And the West Coast is awash
in North Slope oil. Canada is pondering large additional
sales of natural gas to the United States. We don't know
yet what effect new gas-pricing regulations will have on
supply. In fact, new estimates of our own natural gas
reserves have many questioning Federal incentives to switch
to other energy sources.

Oil and gas supply uncertainties are compounded by
questions about energy supplies from alternative sources,
particularly from new technologies that may be economically
feasible by the end of the century. Right now, Congress is
under increasing pressure to subsidize the development of
exotic technologies like extraction of shale oil and liquefaction of coal. But we are uncertain which of the many
choices--including solar and nuclear fusion--will ultimately
supplant oil and gas as our chief energy sources of the
twenty-first century. Certainly, the cost of developing
any of these alternatives overnight (as we managed with the
"Manhattan Plan") is prohibitive in today's economy.

We are concerned about rising oil imports. This anxiety
will diminish as we build up a strategic reserve, as energy
conservation measures take hold, and as alternative domestic
energy sources are called into play. But we will still need
oil imports and there remains the historic political uncertainty in the Middle East. Will a treaty signed in the 1970s
hold for the decades ahead? Can so many nations with so

many economic, religious, and territorial differences keep
a lasting peace--and guarantee a stable flow of oil to the
West? No geopolitical analyst would hazard a serious
prediction.

Raising domestic oil prices to world levels is the one
major step in the President's April 1977 National Energy
Plan proposals which the Congress has so far refused to
take. The current difference between U.S. and world oil
prices surely cannot prevail for the indefinite future. Low
U.S. prices discourage conservation, maintain the demand for
imports, and are at odds with commitments to our allies.

The President took the offensive by proposing to raise
the price of all domestic oil to world levels--via an excise
tax. No issue in recent years has set the Congress to such
bitter squabbling--producer vs. consumer, Southwest vs. the
Northeast, conservative vs. liberal (unless they team up
against the moderates). Like the nation it represents,
Congress is confused and less than certain that the President's war must be waged with the intensity that he asks.
I do believe, however, that the next Congress will come to
grips with the oil pricing issue.

The uncertainty of congressional action has contributed
to the weakening of the dollar. Our trading partners are
understandably concerned about America's resolve to act
responsibly. Forcing U.S. consumers to pay the world price
for their oil and gas will toughen our image abroad; it will
back President Carter's pledge in Bonn. But can any single
measure fully shore up the dollar in such a fragile international economy?

Despite the great uncertainties surrounding any consideration of future energy policy we are, as a nation, maturing
in our response to the problems we face. Greater recognition
of uncertainty and complexity is in itself health. We are
no longer basing policy considerations on simplistic pre-
scriptions such as energy self-sufficiency or massive
investments in alternative energy sources at any cost.
Greater recognition of our economic interdependence with
other nations also tempers our policy responses. We know
that we are not, in one year or two, going to settle upon
an energy policy that will carry us through to the end of
the century.

The Honorable Al Ullman
(Congressman, Oregon)
Chairman, Ways and Means Committee

x

FACING THE INTERNATIONAL ENERGY PROBLEM, 1980-2000

PANEL I

SHARING THE PROBLEMS AND OPPORTUNITIES
OF THE ENERGY FUTURE

The following is a summary of views by an
internationally recognized group of policy
experts with both academic and public service
credentials: Congressman Clarence J. Brown
(R., Ohio), Ranking Minority Member of the
Subcommittee on Energy and Power of the House
Interstate and Foreign Commerce Committee was
in the chair; the panel included Marshall Crowe,
former Chairman of the Canadian Energy Board;
Toyoaki Ikuta, President of the Institute of
Energy Economics, Tokyo; Ulf Lantzke, Executive
Director of the International Energy Agency;
and Carroll Wilson, Director of the Workshop on
Alternative Energy Strategies (WAES), Massachu-
setts Institute of Technology

SHARING THE PROBLEMS AND OPPORTUNITIES OF THE ENERGY FUTURE

For the industrialized nations, finding a solution to
present international energy problems means, in large part,
finding alternatives to massive use of imported petroleum.
Alternatives are many, varied, and costly, thus generating a
set of economic and political choices that form the core of
the so-called energy crisis. Alternatives to imported oil
were described during the panel, including domestically
produced oil and gas, coal, various forms of solar power,
geothermal and nuclear energy. Alternatives to continued
dependence on imported petroleum also exist in the form of
conservation measures designed to reduce overall energy consumption. All of these options must be pursued to avoid
serious economic stringencies through the year 2000; there
is no single answer to our energy difficulties.

In the international market, substantially increased
production of oil and gas outside of the Organization of
Petroleum Exporting Countries (OPEC) is possible which, when
combined with OPEC supplies, suggests the continued availability of these fuels until at least the year 2000--although
serious shortages well before are highly likely. Dr. Wilson
cites the findings of the major WAES study done under his
direction at MIT as follows:

> After two years of study, we conclude that world
> oil production is likely to level off as early
> as 1985. Alternative fuels will have to meet
> growing demands. Large investments and long lead
> times are required to produce such fuels on a
> scale big enough to fill a prospective shortage
> of oil. And the major task for the world will be
> to manage this transition to other fuels and
> renewables in the next century.

While such long-range predictions have not been very accurate
over the last few decades, this consensus estimate appears
to be the best available.

In large part, the role to be played by oil in the
industrial democracies over the next two decades depends on
political and economic decisions to be made in two geographical arenas--the Middle East and Soviet Union. Decisions
facing Arab exporters of petroleum include identifying
long-range production rates to maximize their economic
development, in view of the fact that most of them will
have limited natural resources once petroleum reserves are

produced fully. Also, the Middle East exporters must decide
how to use their petroleum resources to the best political
advantage in the international power arena.

The Soviet Union's choices in the next decade will also
be significant, because the size of its internal supply and
demand could have a major impact on international petroleum
availability and prices, especially if it should become
seriously dependent on imported oil. At this juncture, such
a development does not appear likely because the Soviets
have both a political and economic stake in maximizing
production and in continuing to export a modest amount of
petroleum to the East European nations. But the future
Soviet position will depend upon the speed and efficiency
with which its potentially large deposits of oil in Siberia
can be exploited. Most analysts agree that the Soviet
government will try hard to avoid becoming dependent on
foreign sources, inasmuch as energy is so closely related
to growth in national income and to defense capability.
Whether it can, in fact, avoid dependency is the key question.

As a major importer of OPEC petroleum, the United
States is a key factor in the allocation of the world's
energy resources. The International Energy Agency's (IEA)
statement of policy, to which the United States is a signa-
tory, encourages all IEA members to permit domestic energy
prices to rise as an inducement to both conservation and
additional production. While public policy in the United
States has overwhelmingly endorsed conservation in principle,
there seems to be little willingness to take effective
measures to encourage additional production. Domestic
political conflicts between producing and consuming regions
of the country and public unwillingness to accept replacement cost pricing have stalled progress. Complex regulatory
constraints have also further increased U.S. dependence on
foreign oil. If the market-price system were permitted to
function with minimal interference, domestic production would
assuredly climb, but there seems little prospect--for a
variety of reasons--that the market will in fact be
resorted to.

Coal is one alternative that could provide major relief
in the United States, as well as elsewhere in the industrial
world. Several problems, however, limit coal as a substitute
for oil. Chief among these in the United States are environmental concerns. One prime source of environmental problems
is the relatively large amount of sulphur contained in coal.
While some coal deposits have less sulphur content than
others, all coal has considerably more than the oil and
natural gas for which it might be substituted. Theoretically,
the problem is manageable but, as yet, the level of technological development in pollution control equipment does not
permit acquiring an amount of energy from coal equal to its

oil and gas substitutes without unacceptable degradation of
environmental standards. In addition to air pollution,
other environmental problems from coal include mine drainage, strip mine damage, and ash disposal. All of these
must be addressed before coal can be employed in significantly greater quantities. Some analysts also question the
rate at which coal can be burned without creating the
"greenhouse effect." The buildup of carbon dioxide in the
earth's atmosphere could threaten the earth's weather if
coal were burned in record amounts.

Another major obstacle to wider acceptance of coal as
a suitable energy source is instability in coal supplies.
Strikes by coal miners and changes in public policy make
national and international reliance on coal a risky step.
Few nations would like to find themselves as dependent on
coal from the United States as they have been on oil from
OPEC.

Also, potential importers could only justify large
capital investment in coal burning facilities if the stability of international trade relationships were assured.
Potentially, there is a significant international coal trade
to be exploited if such stability can be achieved. One study
suggests that the United States could ultimately export 500
million tons a year, the equivalent of nearly 10 million
barrels of oil a day.

Domestically, there is a question about the feasibility
of the U.S. short-term coal goals, largely about the capability of the U.S. transportation system to handle the
additional tonnage which the Carter Administration called for
in its National Energy Plan. Coal industry leaders believe
that the production goals can be met and that demand by
utilities will probably reach target levels by 1985; the
difficult questions center on non-utility demand and on
transportation.

Nuclear energy, as an effective alternative in the
United States, is caught in a quagmire of regulatory delay,
adverse opinion in small and aggressive sectors of the popu-
lace, and escalating costs. Outside the United States, on
the other hand, progress in development of standard uranium
and breeder reactors is continuing with or without U.S. help.
For most nations, nuclear power appears to be a sensible
intermediate response to unreliable and high-cost petroleum
imports; for Japan, it appears to be about the only alternative.

Within the United States, and to some extent elsewhere,
long lead times for plant construction, delays generated by
concerns over environmental damage, and fears of proliferation of nuclear weapons are delaying--and may kill--further
widespread adoption of nuclear fission energy. At this
point, investors see great risks in a large-scale commitment to nuclear power. Yet, it is estimated that nuclear

energy has a potential of providing about nine percent of
the world's energy by the year 1985, if obstacles and
uncertainties are removed. Under the same assumptions,
by the year 2000, nuclear power could provide 21 percent
of the world's energy, about the equivalent of 43 million
barrels of oil a day.

In the long run, alternatives such as solar, geothermal
and nuclear fusion energy are expected to provide solutions
to the eventual disappearance of oil and gas as fuels.
Coal gasification and liquification are also expected to
ease the transition into the era of nondepletable energy
sources. The attractiveness of coal for conversion into
other fuels comes in part from its great abundance within
the United States. It provides the United States with a
vast energy source for domestic employment and for export
to other industrialized nations. At this point, the cost
of coal gasification is high, but increasingly efficient
processes are being developed, especially in Europe. Once
the technology is perfected, costs should decline. When
it becomes economically viable, transformation of coal into
hydrocarbon liquids will provide a near-substitute for
petroleum.

Eventually, solar energy in a variety of forms appears
to be the ultimate nondepletable energy source. Present
technology suggests numerous approaches to the use of the
sun's radiation. Windmills may prove economically feasible
in some regions in converting solar-induced breezes into
electricity. Steam production from reflected solar rays
and direct conversion of sunlight into electricity by photo
voltaic cell are also prospects. Meanwhile, the use and
storage of heat from collector panels is being revived
for limited purposes; this relatively cheap source of heat
was popular in some sun-belt regions until the very low
price of natural gas undercut its market. Of course, hydroelectric generation will continue to provide a reliable but
limited-in-quantity source of cheap energy.

Investigations into other exotic sources of fuel are
being pursued by both the private sector and several government agencies. While its potential is not yet clear,
biomass conversion will likely provide an economical
alternative fuel in the future. Another high-technology
source of fuel under preliminary investigation is the nuclear
fusion process, but its potential appears to be in the
distant future, beyond the year 2000.

Further energy conservation also holds out some hope
for a partial easing of U.S. dependence on foreign oil
imports. The most efficient mechanism for encouraging
energy conservation is a higher market price for energy.
However, permitting energy prices to rise to market levels
does not now appear to be a politically acceptable policy
in the United States. This is particularly unfortunate,

for the United States is the major energy user in the world.
Clearly, there could be a sizeable reduction in the present
import level of about 9 million barrels of oil per day if
federal energy policy permitted replacement cost pricing,
as exists in almost all other markets. Paradoxically,
U.S. policy actually subsidizes use of imported petroleum!

Of course, more efficient use of energy supplies has
already begun in the United States and other industrial
nations in response to the higher prices for OPEC oil established in 1973-74. The downward shift in the ratio of
growth in energy use to growth in GNP, from over 1.0 to
about 0.6 or 0.7, in the last five years in the United
States demonstrates the effectiveness of market-initiated
incentives. While making predictions of future energy to
growth rates is risky at best, there is good reason to
believe that the 0.6 ratio is not far from a long-term
sustainable level. As additional adjustments to the higher
real price of energy are made over time, new sources of
energy-use efficiency will be reflected in the statistics.

Lack of adequate conservation incentives is one reason
why some nations have not been successful in their efforts
to reduce dependence on foreign supplies. Full cost pricing
is necessary to achieve success through market mechanisms.
In the United States and the United Kingdom, for example,
natural gas prices should be allowed to rise for this
purpose, and in Italy electric power rates need to reflect
the full cost of production.

Artificial attempts are also being made to induce
conservation in industrial countries. The announced U.S.
policy is to require a shift for some uses of oil and gas
to other fuels. Such is the case in the electric power
industry, where the Carter Administration has proposed that
no new generating stations be built that would burn oil or
gas and that after 1990 no utility be permitted to burn
these fuels. In the automobile industry efficiency
standards enforced by government regulations are the
current answer to conservation in highway travel. In
addition, proposed U.S. policy includes tax incentives for
increasing energy conservation at home and in the factory,
as well as for installation of equipment for using other
forms of energy as a method of conserving gas and oil.

Outside of the United States, the degree to which
conservation can help ease the energy crisis varies among
countries. In Japan for example, fuel efficiency of
industrial processes is already high and substantial additional conservation appears to be impractical given the
present level of technology. Japanese investments in energy
conservation equipment are already producing returns. At
the other side of the spectrum is Canada, as well as the

United States, with great potential for conservation. More
conservation will be evident in these countries, over time,
as energy users have an opportunity to substitute additional
capital, technology, or labor for energy input. Worldwide
it is estimated that the potential for conservation is in
the neighborhood of 12-17 percent of all energy usage.

Complicating all issues has been the fact that the
energy price rise generated by the OPEC cartel arrived in
the United States at about the same time that public policy
was beginning to concentrate on improving the environment.
In trying to cope with the OPEC-inflicted embargo and price
increases, public officials have also had to deal with
environmental issues as they considered alternative strategies
designed to reduce oil consumption and to reallocate scarce
fuels to their optimal uses. Thus some of the additional
costs, regulations, and confusions imposed on energy users
were the result of increasingly stringent environmental
standards, as well as steeply rising oil prices.

One characteristic which marks the current debate on
energy, at least in the United States, is the extent to
which government will function as allocator of resources
instead of the market. The market heretofore has been the
traditional allocation mechanism and thus the vehicle by
which transitions from one energy source to another have
been accomplished. Unfortunately for U.S. policy and progress, the definition of the allocation problems now faced
by the nation has become lost in political rhetoric,
bureaucratic linguistics, and regional jousting. Special
interest groups in private and public sectors have launched
extensive campaigns to protect the status quo or to
establish footholds in the new energy environment.

One factor that makes government initiatives more
acceptable to citizens of the industrialized nations is
the visibility of the energy "crisis." Centralized action
was much more acceptable during the early days of the price
increase when people believed that market mechanisms were
ineffective. The rapid growth of national and international
energy agencies during 1974 is evidence of this phenomenon.
Since then, substantial increases in petroleum production
outside of the OPEC cartel, plus market-induced conservation and economic slowdown, have led to a temporary halt
in the rise of the international price of oil--indeed, real
prices have been declining for some time. The petroleum
"glut" has reduced the crisis atmosphere, with the result
that governments find that achieving cooperation from their
citizens and from each other is more difficult.

Perhaps the most difficult aspect of managing the
energy transition, characterized by increasing prices and
the possibility of another oil embargo, is the development

of cooperation between governments. Shortly after the
embargo of 1973-74, the International Energy Agency (IEA)
was formed by Western industrialized nations to formulate
strategy for international cooperation. To ease the
impact of another massive shortage, which could arise out
of another 1973 type embargo, an emergency sharing agreement
was created to ensure equitable allocation of any shortfall.
Incentive for such cooperation arises from recognition by
most nations of their economic interdependence. During
the 1973 embargo, for example, West Germany supported the
Dutch when OPEC terminated supplies to the Netherlands.
Without petroleum, the Dutch would be unable to supply
food for West Germany.

The allocation system established by the IEA is tested
once every two years to assure its workability. This exercise performs another function as well, namely, it demonstrates the capability of this system to the world. In the
words of IEA Executive Director Ulf Lantzke, "The mere
existence of the emergency system is an additional element
preventing a crisis."

On a broader front, IEA members are developing parallel
national programs to reduce dependence on imported petroleum.
As already noted, such long-term cooperation can be much more
difficult without a crisis atmosphere. In October 1977,
nevertheless, an agreement was reached by IEA members on a
long-range strategy committing each member-nation to adopt
twelve principles in formulating its national energy policies.

Among the principles established by the IEA agreement
are reduction of oil imports by conservation and substitution
and allowing domestic energy prices to rise in order to
stimulate conservation and to encourage additional supplies.
Emphasis is placed, too, on decoupling rates of energy demand
growth from overall national income growth. Prior to the
recent price increases, these two growth rates have tended
to be almost identical. Encouragingly, energy growth rates
for 1977 are well below national income growth rates, but
it is premature to suggest that this problem has been
permanently settled.

Substitute forms of energy receive heavy emphasis in
the agreement. Addressing the problem of Japan and Western
Europe, the document calls for replacement of fuel oil as
a source of electric power generation with such other sources
as nuclear fuel and coal. The agreement urges a commitment
to nuclear development, although two constraints--public
opinion and investment risk--are expected to hamper progress. Several nations could not endorse this concept due
to their own domestic policies; however, all of them
endorsed the policy in principle. In recognition that
nuclear energy is only a part of the near-term answer, IEA
has endorsed expanded international trade in coal. The long

lead time necessary for mine, plant, and transportation construction requires immediate action if this initiative is to have some impact by the mid to late 1980s.

The 1977 IEA agreement urges all governments to make long-range plans for energy production and consumption. This includes settling issues concerning environmental protection and regulatory control in order to provide a favorable climate for investment. It urges cooperation in research and development of new technologies. And finally, it asks for close cooperation with developing countries and a better dialogue with producing nations.

The urgency of accomplishing the goals described in IEA agreements can be appreciated by reviewing future world oil production and demand statistics calculated by various international research projects. Setting aside the oil requirements of Communist nations, the Workshop on Alternative Energy Strategies estimates world demand for all forms of energy at between 160 and 210 million barrels of oil equivalent per day (MBDOE) in the year 2000. Of this total, oil demand is expected to be between 75 and 93 MBDOE. Production of non-Communist world petroleum--now at 46 MBDOE--could meet this demand only if a number of highly optimistic assumptions are realized. More realistically, severe shortages are likely by the late 1980s or early 1990s. Of course, the supply-demand squeeze could begin much earlier should the Saudis, for example, decide to hold rather than expand their present production rates. In order to avoid a massive constraint on the world's economic growth potential in the late 1980s or 1990s, it will probably be necessary for the Arabian peninsula producers not only to avoid cutting their production but also to nearly double production, to about 25 million barrels a day.

Compounding the problem of international cooperation is the great diversity found among the industrial democracies in degree of dependence on energy imports. Canada, for example, is near complete independence; it does import some petroleum from the Middle East and Venezuela, but it is a small net exporter of oil--primarily to the United States. Canadian tax policy has been successful in encouraging a great deal of exploration and development of gas and oil supplies. With additional opportunities in tar sand production, Canada can continue to contribute toward solving international energy problems. Japan finds itself on the opposite end of the scale with respect to dependence on imports. Exacerbating its almost total dependence is the fact that it has already pushed conservation to its limit without interfering with economic growth. In Japan, expanded use of coal is imperative, but this too continues the country's dependence on imports of energy and produces additional problems such as air pollution and coal ash

disposal which have not yet been solved. Nuclear power is
another option for Japan, but there, too, there are
difficulties--largely because of U.S. pressure regarding
its nuclear proliferation worries. Thus, Japan is in the
forefront of the emerging era, facing problems and uncertainties that have not yet completely thrust themselves
on the rest of the industrialized nations.

   Of all the variables confronting energy decision-
makers, the uncertainty about future governmental policies
is perhaps the most damaging if incorrectly handled.
"Political" risk exists in the present energy market on
both the producer and consumer sides. Quadrupling of
international oil prices, for example, was a move that
could only be accomplished by a government-backed cartel.
Governments of consumer nations have vacillated on energy
policy, depressing investment directed toward the very
solutions needed by consuming nations. Risks associated
with energy investment are further intensified because the
world is experiencing both a transition and a rapid expansion in energy consumption. Some idea of the risks involved
can be gained from the observation of one expert commenting
on estimates of the future:

> I am pessimistic about the accuracy of any forecast
> and do not know of any forecasts of twenty years
> ago that are worth anything today. I don't seri-
> ously believe that even the predictions made in
> the WAES study will turn out in the way that we
> forecast.

   One area where uncertainties need to be ironed out is
in governmental policies. In the United States, environmental protection policy is a case in point. Clear air
standards are increasingly stringent as new laws are enacted
by Congress. And, even if new legislation is not expected,
existing statutes often leave enforcement standards to the
discretion of regulatory agencies. It is often necessary
to await a case-by-case approval for investments in projects
with obvious environmental impact.

   As uncertain as specific levels of supply and demand
or government regulation may be for the future, one thing is
clear: the cost of energy will be increasing. Every contribution from whatever energy source will be necessary to
avoid dramatic increases. Since the latter 1980s may see
the start of the next crisis, national energy policies--
including international cooperation--must be established
without delay. Taking account of essential lead times,
1985 is already yesterday.

*   *   *   *   *   *   *

Summary Statement by the
Honorable Clarence J. Brown

As this report on "Sharing the Problems and Opportunities of the Energy Future" makes clear, the problems besetting a national energy program are not being addressed as effectively by the federal government as one might hope. Instead, those trends which have helped to ameliorate the crises of recent years are those which have been generated by the market-price system. Rising prices have produced an increased awareness of the need to conserve all types of energy in homes, transportation, and industry, and the ratio of growth in energy to growth in Gross National Product has fallen in response to the increased relative price of energy. At the same time, alternative sources of energy have attained new heights of economic feasibility due to cartel level pricing of OPEC oil. Solar heating, nuclear generation, and other more exotic energy systems now appear more attractive.

Despite a recognition by almost all national leaders of the overriding public interest in reducing dependence on foreign energy supplies, no genuinely constructive national policy has been created to reduce that dependence. In fact, present policy actually encourages further foreign dependence by failing to encourage conservation and by snarling the constructive efforts of energy producers in costly regulatory red tape, resulting in higher prices to consumers with little net increase in energy supplies. The refiners entitlements program provides a good example. The more foreign petroleum which a refiner imports, the greater the subsidy paid to the firm, thereby subsidizing energy dependence.

There is a great deal of contradiction in our present energy policy. While our stated goal is to reduce dependence on foreign energy, federally administered price controls prevent replacement cost pricing. Moreover, we are encouraging importation of substitutes for domestic gas at prices far above the domestic price. The result is less gas at a higher cost to consumers, and significantly, a continuation or exaggeration of national dependence on insecure imports.

The National Energy Act recently signed into law is supposed to relieve U.S. dependence on imported oil and to correct international financial distortions. Yet the new law is no improvement over the policies of recent years. Natural gas is still regulated. In fact, price controls are extended to the intra-state market for the first time. Only a small amount of gas is actually deregulated. In addition, the incremental pricing provisions of the new law create further distortions by forcing industrial users to pay a higher than average price for gas and "hiding" the cost of gas to the final consumer.

12

Regrettably, public policy has yet to develop an effective solution to a costly national dependence on foreign energy supplies. Instead, recent tinkering by the federal government has only produced a rapid multiplication of regulations leading to an increased cost to customers. Replacement cost pricing for energy, now commonplace in other industrial democracies, will have to become our national policy before the energy crisis is solved.

PANEL II

FROM WHERE AND HOW CAN THE
WORLD'S OIL NEEDS BE MET?

The session of the Fourth Quadrangular Conference
on the issue, "From Where and How Can the World's
Oil Needs Be Met?" was chaired by Nathaniel Samuels,
Vice Chairman of Kuhn Loeb Lehman Brothers Inter-
national; Chairman, Louis Dreyfus Holding Company,
Inc.; and former Deputy Under Secretary of State
for Economic Affairs. Panel members included
Maurice Granville, Chairman and Chief Executive
Officer, Texaco; Senator Mike Gravel of Alaska,
Chairman of the Subcommittee on Energy and Founda-
tions of the Senate Finance Committee; Dr. Ulf
Lantzke, Executive Director of the International
Energy Agency; and Dr. John H. Lichtblau, President,
Petroleum Industry Research Associates, Inc.

FROM WHERE AND HOW CAN THE WORLD'S OIL NEEDS BE MET?

     The panel discussed the implications of the following
questions: will there be an oil crisis? Will the years
1985-1995 find the world with an uncovered demand for
energy? Can this situation be avoided and, if so, how?
What is the government's role in ensuring a continued
adequate flow of oil while transition to other energy
sources is progressing? The discussion reflected considerable uncertainty regarding the supply and demand
sides of the equation.
     On the supply side, no one can accurately predict the
success that will be achieved in discovering new sources of
oil or the total exploitable quantities of known, but as yet
relatively untapped, sources in Mexico, the Canadian Arctic,
and elsewhere. Moreover, unforeseen political developments
in Saudi Arabia or other major oil producing nations can
have significant impact on oil production and supplies. In
considering energy supplies in general, much will also depend
on how rapidly and well alternatives like nuclear energy and
coal are exploited. If there are serious time lags in their
development, then demand for energy must be covered by oil,
forcing a more rapid rate of use and depletion.
     Future demand for oil is equally difficult to predict.
The number of factors influencing demand are myriad and it
is difficult to determine whether there will or will not
be a "crunch" in the 1980s or 1990s. Projections can be
marshalled to support either side of the argument. At
present there are factors working both for and against a
shortage. Those that would indicate a shortage is not
likely are the slower rate of economic growth in the
industrial countries of the West and the increased efficiency in the use of energy per unit of GNP. A factor
working for a shortage is the present glut of oil, for
the present easy availability may dilute incentives to
develop the needed new oil sources. The OPEC price of oil
has been declining in real terms in the last fifteen
months, as a result of the glut in part, and may also
serve as disincentive to developing other oil sources.
What the balance between these forces will be in the
future is impossible to determine at the present time.
     Over the next decade or two, the problem is not
essentially one of physical availability of energy, but
rather one of "timely availability." Government policy
must be designed to meet the long lead-times necessary to
bring new oil on stream and to develop other energy sources
without creating periods of shortage. One expert summarized
the argument:

> [An energy crisis] is not inevitable.... [Meeting
> the challenge] requires market factors, market
> prices, and intelligent government policy. We
> can avoid a dangerous situation from ever develop-
> ing, because essentially the amount of energy
> available in the world is enormous. There is no
> way you could have a permanent energy crisis; the
> worst that could happen is a transition period
> during which not enough is available and, as a
> result, prices would soar. But this can be avoided
> both on the demand side and on the supply side.
> Whether the policy will be the right one--that is
> the question!

Scenarios regarding the future of oil depend, to a large
degree, on assumptions about the rate of economic growth
and the price of energy. Recent IEA estimates were cited,
indicating that, if one assumes a rate of economic growth
of 4.4 percent and a growth rate of oil consumption of 2.9
percent in OECD countries and 3.5 percent in the rest of
the world, OECD countries would import 28 million barrels
per day (mbd) in 1980 and 32 mbd in 1983. The Organization
of Petroleum Exporting Countries (OPEC) would have to produce more than 35 mbd to meet world demand by 1985. These
figures are probably a bit too pessimistic, since there
is likely to be less average economic growth than assumed.
The conclusion from this is that there will likely not be
any real difficulty in the oil market until at least 1985
and probably later. If one assumes a 2 percent increase
in energy demand, then a shortage is unlikely. If, however,
the rate of increase is 4 percent, then the situation is
reversed and the possibility of a shortage is dramatically
heightened. A recent study by a European oil company showed
Europe's demand for oil growing at a rate of only 1 percent.
U.S. demand growth has also slowed significantly. The
demand for oil, then, is growing at a lower rate than that
for other sources of energy. Such a trend provides grounds
for optimism.

The most important factor influencing demand for oil
will be price. Supply and demand will always balance at
some price. At one price, a certain amount of oil will be
obtainable; at a higher price, more will be supplied and
less demanded. It is not a linear relationship, to be
sure, but they are related. If the price of oil had not
risen drastically in 1973-74, for example, neither the
finds in the North Sea nor those in Alaska would have been
economically viable. Mexico, too, was strongly encouraged
by the 1973-74 price increase. As late as 1974, it was a
net importer of oil. Given the price rise, however, the
government decided to allocate greater capital to development

of its own resources. Although its potential is at this
juncture not clearly defined, Mexico is expected to become
a major oil producer in the future.*

Even in the more distant future, oil will be available
at some price. New resources will be tapped, however, only
if the price of energy increases to make exploitation of
those resources economically viable. Only then will crude
oil with heavier densities and greater sulphur content be
fed into redesigned refineries. Only then will deeper wells
be drilled in less hospitable areas.

The danger, however, is that the price of oil will
become so high that it will have a negative impact on the
world economy and induce economic stagnation. In 1973-74,
the jump in oil prices, particularly in light of the short
period in which the fourfold increase occurred, shocked the
economies of the world, reduced national income, created
unemployment and seriously aggravated an already rising
rate of inflation.

One reason the impact of the price increase in 1973-74
was so severe was because the price of oil had been kept
artifically below replacement costs for an extended period
of time. Oil prices must reflect the cost of the next barrel
of oil, which, almost inevitably today, will be more costly
than the previous one. If national policies are designed to
continue to stifle price increases so that consumers are
paying a price of energy below replacement costs, then
future problems will be intensified. Since someone must
pay for the cost of replacing depleted energy resources,
such a policy misleads the public and inhibits conservation
efforts, as well as depresses supply. It also delays the
day when adjustments to increased energy costs must arrive
and, therefore, increases the difficulty of the inevitable
transition to higher cost energy sources.

On the supply side of the equation, future policies of
the key producing states, and Saudi Arabia in particular,
will be a major factor in determining what oil will be
available in 1985 and beyond. Forecasts suggest that by
1985 member countries of OPEC will contribute between 55-60
percent of the world's oil supply. Moreover, the Persian
Gulf region will represent 50 percent of OPEC's total
productive capacity in 1985. Saudi Arabia will be the
dominant Persian Gulf producer accounting for 30-40 percent
of the Gulf's total production capabilities. Beyond 1985,
Saudi Arabia's dominance will become even more considerable
unless immense new fields are discovered.

It has been estimated that OPEC oil production will
have to rise to between 35 and 40 mbd, of which Saudi

---

*A CSIS monograph on *Mexico's Oil: Trends and Prospects
to 1985*, by Sevinc Carlson, was on the table at the
conference. (See Appendix.)

Arabia's share will have to be at least 15 to 16 mbd, in
order to meet world demand in 1985. The Saudis, however,
have signalled that they do not perceive any urgent need
to expand production capacity so rapidly. Given the
declining value of world currencies and the need to find
investments that are politically practical, the Saudis
feel it may be in their best interest to leave their
oil in the ground. In short, they have indicated their
production will not reach 15-16 mbd by 1985. U.S. energy
planning, however, has proceeded on the assumption that
it will. This disjuncture in planning could be dangerous;
accordingly, the United States should revise or at least
hedge its assumptions.

Saudi Arabia's decisions regarding its future oil
production will, of course, be influenced by political
factors as well as economic considerations. The Saudis
cannot isolate themselves from their regional and international political setting. They will not be immune,
for example, from forces at work in the Persian Gulf,
including national rivalries and radical elements pres-
suring for changes in the Gulf's conservative regimes.
The Saudis also have an active interest in resolution of
the Arab-Israeli conflict. It is very likely that if
another Middle East war were to erupt, Saudi Arabia would
again be involved in an oil boycott.

Domestic upheaval would also seriously impact on Saudi
Arabia and other Gulf producers. The assumption of power
by radical forces less friendly to the West than present
regimes in the Gulf area could result in decisions to cut
production and raise prices. Internal conflict stemming
from modernization pressures may spill over into several
of the major oil-producing regions. Any number of scenarios
could be devised regarding domestic developments in Saudi
Arabia, for example, the end result of which could be Saudi
oil production at levels well below those that have been
predicted. Although American energy policy cannot be
formulated on the basis of such uncertainties, it must
contain sufficient flexibility to adapt to various changes
in the supply of oil.

While availability of oil over the next two decades
depends greatly on policies of, and political developments
in, current oil producers like Saudi Arabia, availability
will also be influenced strongly by the success achieved in
tapping new oil sources. With respect to known new sources,
prospects are mixed. North Sea production, for example,
could double but would still not make a major dent in world
needs. On the other hand, Mexico could make a long-range
difference; its reserves are now estimated at about 20
billion barrels, and could be as many as 50 to 60 billion
barrels. Production in the McKenzie Delta of the Canadian
Arctic could also be significant, but not until at least the

1990s. Further prospecting in the United States has not
generated great excitement; moreover, little can be
expected, in the short to medium-term at least, from the
more exotic sources of oil such as shale.

A particularly interesting prospect with great potential
is the Orinoco petroleum belt. Venezuela's opportunity in
this area is a good example of potentials and problems
inherent in many newer oil sources. It is estimated that
the Orinoco belt, some 600 kilometers long and 60 kilometers
wide, could possibly yield as much as four trillion barrels
of oil. Only the eastern third of the belt has been partially
explored, and it has been determined that there are enormous
reserves there. The Orinoco belt will provide some lighter
oil, considerable amounts of heavy oil, and enormous quantities of extra-heavy crude.

There are two major problems, however, confronting
exploitation of petroleum in the Orinoco area: first, there
is little knowledge of reservoir behavior; second, Venezuela
faces a stiff challenge in upgrading the crude. To counter
these problems, a vast development program has been initiated.
Between 300 and 400 wells will be drilled to determine the
more interesting areas of the belt. Moreover, up to 10
smaller pilot projects will be undertaken before the first
major pilot project.

It is hoped that oil from the Orinoco belt will begin
coming on-stream within ten years. Resources from other
discoveries also will likely become available after a decade
or so; but will they be able to offset a tightening market
situation? This is the key question in the current energy
debate. One expert concisely summarized the problem:

> We just don't see what additional reserves could
> be made available from new major discoveries.
> The existence of Alaska was known about 10 to 12
> years ago, Mexico a bit later. The lead times to
> bring new oil to market are such that we would
> have to know today about new major oil discoveries
> so that by 1985 we could again be in the happy
> situation in which a rather tight market situation
> could be offset by new oil coming on-stream, much
> as we are experiencing today.

Furthermore, it was argued that adequate oil supplies
can be made available in the late 1980s and 1990s only if a
climate is created today that encourages conservation and
stimulates exploration in costly and riskier areas. The
government is, to a large extent, responsible for creating
this environment and, to date, it has done little to fulfill
that responsibility.

Indeed, a great part of the present energy problem can
be best expressed as a "crisis of energy policy." One problem

has been inconsistency. The U.S. Government, for example,
has held down oil and gas prices artificially, encouraging
excessive use of these fuels by the American people, and
then has stepped in with costly regulations to promote
conservation.

An additional difficulty generated by government policy
has been continuing uncertainty in the energy sector.
Natural gas deregulation is one example. For over 20 years,
market participants have been uncertain whether or not
natural gas would be deregulated. They have been kept
guessing as to whether sufficient supplies of natural gas
would continue to be forthcoming. Consequently, planning--
for equipment to transport and consume gas, for example--
was made extremely complicated. Environmental, job safety,
and licensing regulations have also been changed so fre-
quently that many capital investors have either abandoned
some projects or increased their required rates of return
due to increased risks.

One expert pointed out that the government's policy
problems are, essentially, a function of demands of the U.S.
electorate, and if there is a problem, "it rests with the
people." They have expected to get their energy "on the
cheap," below replacement cost. Thus far, at least, the
majority of their representatives have responded to this
expectation. Eventually, however, a price will have to be
paid. Already a balancing effect is in operation in the
form of a "muddling" of the American economic situation.
The enormous balance-of-payments deficit derived from huge
quantities of oil imported to meet the insatiable American
demand for energy is but one example. One political leader
concluded:

> Only when there is sufficient pain among the
> American electorate, the American consumer, will
> he get knowledgeable about the workings of the
> economic marketplace and prepare himself to pay
> the proper price of oil and gas.

If problems in the energy policy of the U.S. Government
are grounded in expectations and demands of the American
electorate, they are most clearly manifested in relationships between government and the oil industry. From
industry perspective, at least, the two have not always
cooperated and sometimes they have found themselves at
each other's throats. In the words of one expert:

> The industry (and the private sector in general)
> can respond to policy initiatives and market changes
> when relatively free to do so, and it can respond
> efficiently. If it is to do so, the industry has
> a case for expecting government recognition and

understanding corresponding to the careful
understanding the industry accords to government.

Unfortunately, mutual understanding has not always been
the case. Government has not always recognized and under-
stood industry problems; nor has it left industry free to
respond to the changing international energy environment.
In fact, government regulation of industry has placed numerous
constraints on oil production. Of all the economic sectors,
it was pointed out, both the greatest degree of governmental
control and the most serious national problems lie in the
energy sector. Is there a lesson to be drawn from this?
While constraining production through regulation, the
government has done little to generate positive incentives
in the search for and exploitation of new energy sources.
It could be argued that it has provided disincentives through
regulation. Furthermore, deregulation of natural gas would
likely cause a higher finding rate, and while the exact
quantity cannot be determined, it would not be an insignificant amount. Deregulation would also promote the search for
deep natural gas in the United States which has, thus far,
been neglected.
The problem of deregulation is essentially a price
problem, and the government's policy regarding the price
of energy is, perhaps, the greatest disincentive facing the
oil industry. The United States has kept energy prices
below replacement cost. Such a policy cannot continue. It
was suggested that the government should now raise the price
of energy to the world level. Unlike an oil users tax that
would decrease oil imports in the short-term and have only
a marginal impact, raising the price of energy to the world
level would, over the long term, induce a decreased demand
for imported oil. Moreover, it would stimulate conservation,
provide incentives for further exploration and strengthen
the international position of the dollar. "Excess profits"
by the oil companies could be recovered through taxation.
Reservations were expressed over yet another aspect for
U.S. policy, namely, "the efficiency of the mechanism for
apportioning energy to society." Government has not given
as much thought to this dimension of the problem as to others,
but its impact on the United States could be as serious as
a real lack of energy supplies. If the apportioning
mechanism is not efficient, it will produce economic dislocations and monetary problems leading to a national,
and perhaps international, recession.
It was clear to all that the future of oil and gas is
uncertain and governmental action is now required. Even if
there is no shortage of these fossil fuels in the latter half
of the 1980s, government will still have to play an active
role in planning the inevitable and extremely important
transition to other forms of energy. The nature of

transition is vital to the future energy status of this
country and the world. It is only through sound planning
by both government and industry that it can be made orderly
and effective. The United States will continue to depend
on oil and gas until that transition can be made, and if
a shortage is to be avoided in these areas, planning by
both sectors will have to proceed along well constructed
lines.

Delay cannot be afforded. In order to meet demands
imposed by transition from oil and gas, and to ensure
sufficient supply of these fuels during the transition
period, immediate action is required. Unfortunately, the
United States has been without an energy policy for 18 months
and unless such a policy is soon formulated, the opportunity
to confront effectively problems that lie ahead may be missed.

By the second half of the 1980s, the countries of the
world, and the United States in particular, could reach a
point at which they suffer considerable economic and political dislocations due to a tight supply and demand situation
in the energy sector. Panel members agreed that several
options could be pursued in order to avoid this situation.
Some of those options, such as sacrificing economic growth,
are clearly unacceptable. Others, such as the promotion of
conservation measures like the German program of incentives
to industry, can help a great deal. In the final analysis,
the challenge can only be met by a program grounded in a
wise government policy. It is not at all clear that such a
policy, at least in the United States, will be forthcoming.

*   *   *   *   *   *   *

Summary Statement by the
Honorable Nathaniel Samuels

The foregoing interpretive report summarizes well the
character of the panel's discussion and its main findings.
My own conclusions from those discussions can be stated as
follows:

(a) We must use market forces and the resources of
the private sector to the maximum extent if we are
to meet the oil supply problem.

(b) Government and private sectors must work together closely to ensure a sensible, timely
transition from high dependence on vulnerable
oil imports.

24

(c) The West needs an "oil diplomacy," a comprehensive approach that adequately takes into
account the impact of petroleum on the world
monetary situation, the geopolitical position of
the principal producers and consumers, and the
international politics of oil and petrodollar
transfers.

PANEL III

THE INTERNATIONAL POLITICAL DIMENSIONS
OF NUCLEAR ENERGY

The third panel of the Fourth Quadrangular
Conference on "The International Political Dimen-
sions of Nuclear Energy" was chaired by Representa-
tive Mike McCormack (D, Washington), Chairman of
the Subcommittee on Advanced Energy Technologies
and Energy Conservation Research, Development, and
Demonstration, House Committee on Science and
Technology. Members of the panel included: Karl
Kaiser, Director of the Research Institute of the
German Society for Foreign Affairs in Bonn; Joseph
S. Nye, Deputy Under Secretary of State for Security
Assistance, U.S. Department of State; Keichi Oshima,
Department of Nuclear Engineering of Tokyo
University; Henry S. Rowen, Professor of Public
Management, Stanford University; and Mason Willrich,
Director of International Relations for the
Rockefeller Foundation.

INTERNATIONAL POLITICAL DIMENSIONS OF NUCLEAR ENERGY

The international politics of nuclear energy are now undergoing a fundamental change with very important conse- quences for the rules and institutions of the present system, and, consequently, for the behavior of states within the system. Furthermore, the nature of the problems facing government policymakers and industry in the exploration differ, enlivening the debate, while obscuring the quick solution.

Decisions now being made and implemented will have their impact in the 1980s. There is, therefore, an urgent need for a new international consensus on the role nuclear energy should and will play in the years ahead. We must start from where we are; a workable regime must be created from the existing nuclear power industry and established international nuclear relationships. While conditions of national sover- eignty and an unequal distribution of resources will restrict what can be made workable internationally in the nuclear field, a maximum effort must be made. Finally, it must be realized that the role of nuclear energy, while necessary, is and will be limited (primarily to the generation of electricity) for some time to come. Nuclear energy is not a panacea for world energy ills; it is probably not even economically viable for many countries. In short, a new international understanding about nuclear energy must be developed if it is to be exploited effectively and undue problems are to be avoided.

In part, the need for a new international consensus regarding nuclear energy stems from the fact that the con- sensus that existed in the 1960s and early 1970s has broken down. Three reasons were suggested:

First, after the 1973-74 oil crisis, assured access to energy supplies came to be perceived by most countries as a vital requirement. Dependence would not only cause higher prices for energy, it would also represent a measure of political and economic leverage that could be wielded by some countries over others. Nuclear energy is seen as a means of achieving national energy self-sufficiency; accord- ingly, it has acquired an appeal and legitimacy previously not enjoyed.

Second, India's 1974 peaceful nuclear explosion (PNE) demonstrated that nuclear energy had spread to the Third World and thereby to areas with traditional and very difficult conflicts.

Third, sales of sensitive technologies by the Federal Republic of Germany and France to Third World nations raised questions in some minds about the adequacy of the existing rules of the international nuclear power system.

Several differences have emerged in recent years between
the United States and its West European and Japanese allies
regarding how best to proceed with development of nuclear
energy. They stem from different national experiences and
priorities as well as differences in energy endowments. One
disputed issue is over the nature of nuclear safeguards. In
the past, safeguards were considered a device; first, to
indicate misuse of nuclear materials and, then, to alert the
international community to that misuse. Today, some
policymakers--Europeans and Japanese consider them to be
mostly American--are trying to reorient the meaning of
safeguards to make them a device not merely to indicate
misuse, but to prevent it.

A more important divergence in European and American
approaches exists on the question of transfers of sensitive
nuclear technologies to Third World countries. One participant pointed out that the problem is perhaps not really one
of "sensitive technologies" but of "sensitive countries."
The United States strongly opposed French-Pakistani and West
German-Brazilian nuclear deals largely because of the Third
World involvements. At the same time, the United States was
objecting, it was negotiating an agreement with Japan about
reprocessing on an experimental basis. If it were technology,
and not the countries involved, that was "sensitive," then
a U.S. deal with Japan should have been opposed as well.

Differences in European and American views of this
issue can be seen especially clearly in the West Germany-
Brazilian nuclear deal. The agreement--signed in June 1975
and worth approximately $5 billion--opened the way for a
large number of commercial contracts between West Germany
and Brazilian firms which are intended eventually to provide
Brazil with an autonomous nuclear industry. The business
arrangements are complex, calling for uranium prospecting,
mining and conversion to begin immediately, along with
construction by the West German firm, Kraftwerk, of two
nuclear reactors of the 1250-megawatts pressurized-water
varieties. Each is expected to take six years to complete.
In addition, a heavy-components facility is to be built in
Brazil by a West German-Brazilian consortium, and a pilot
fuel-element fabrication plant is being planned by Nuclebras,
the Brazilian state utility, with two West German concerns.
Finally, a uranium enrichment facility and reprocessing
plant are to be constructed. It was the enrichment
facility, in particular, that generated controversy, in
part because it was to be based on the as yet commercially
unproved "jet nozzle" process.

From the German perspective, accusations of reckless
commercialism emanating from some quarters in the United
States were totally unwarranted. It was pointed out that
the bargain between the two countries is completely

consistent with international agreements. Moreover, benefits
will be derived for the entire international system, not
just the Federal Republic. For example, through the deal,
Brazil agreed to a system of international controls, "no
minor accomplishment." In addition, as part of the deal,
Brazil is the first country to accept the principle of
international storage of plutonium.

In general, the Federal Republic and France--among
others in the London Suppliers' Club--believe that if denial
of nuclear technologies to Third World countries is pushed
too far, some countries will quit the "international nuclear
control system" completely. They will decide to pay the
price necessary to obtain those technologies, with no
controls, threatening a breakdown of the entire nonproliferation system.

A third issue on which no consensus now exists is the
question of disposition of spent nuclear fuels. Current U.S.
policy calls for retention of a U.S. veto over disposition
by other countries of spent fuels originally supplied to
them by the United States. Such an option is politically
unacceptable to some countries. Others would want U.S.
rights more clearly defined before accepting the veto.

Serious as these differences are, they pale in comparison with divergence in American and West European-Japanese
views of the breeder reactor and reprocessing and enrichment technologies. U.S. policy supports adoption of the
once-through fuel cycle, preferably based on the light-water
reactor. It also calls for indefinite deferral of reprocessing by the United States and others, plus delay and
restructuring of U.S. and other breeder reactor programs.
These calls are made in order to emphasize technologies
more proliferation-proof than the liquid-metal fast breeder
reactor.

To Europeans and Japanese, this complex of policies is
unacceptable. They are convinced of the long-term need
for breeders and reprocessing. Moreover, the "once-through"
process based on light-water reactors would prolong their
one-way dependence on the United States for enriched fuel.
Further, it would continue a dependence on only a few uranium
suppliers--Canada, South Africa, and Australia. In the
long-term, security of these supplies could not be guaranteed.

Western Europe and Japan regard reprocessing and
breeders, then, as essential for reasons of economy, ecology,
and, most importantly, security of supply. From the
Japanese perspective, in particular, no other viable
alternative exists, except oil imports. Such an option,
however, only serves to heighten Japan's dependence when
energy self-sufficiency is the desired goal.

The Japanese and West European emphasis on breeders
and reprocessing underlines the priority that security of
supply considerations have assumed. It reflects their view

that reliance on outsiders for energy supplies is a risky
policy. In general, too, Europeans and Japanese are
convinced that reprocessing and breeders are a rational
and safe way to achieve waste disposal and stretch resources.
On this latter point, the fast breeder reactor constitutes
a difference in quality from other nuclear processes. He
cited the following figures: A light-water reactor produces
5,000 megawatt-days per ton, while a heavy-water reactor
produces 10,000-20,000 megawatt-days per ton. In contrast,
a fast breeder reactor will produce 500,000 to 800,000
megawatt-days per ton. The difference in the order of
magnitude of these figures, he implied, was a clear reason
for opting for breeders.

Given the long-term outlook and lead-times involved in
developing energy resources, neither Europeans nor Japanese
share the American view that there is time to delay breeder
programs. The United States might be able to delay because
it is relatively rich in energy resources. Europe and
Japan, however, are resource poor and cannot afford the
luxury of time. In fact, there is a strong perception in
Europe and Japan that they are running out of time. It was
also mentioned that if the Europeans delay breeder reactor
programs, they could lose the technological lead over the
United States they presently enjoy.

In proposing a delay of breeder reactor programs, the
U.S. Government is aware of the energy problem of its allies,
but it argues that time is needed to ensure that adequate
safeguards can be developed and that a breeder reactor
program can be justified economically. In his energy
message of April 1977, President Carter did not oppose
national breeder reactor programs _per_ _se_. He only opposed
the premature commercial use of breeder reactors, arguing
that other alternatives should be pursued first. If, after
alternatives are explored, some nations still decide to
pursue a national breeder program, the Administration would
not oppose them so long as the countries involved assumed
responsibility for "externalities" of their programs, that
is, potential costs placed on others and potential security
costs.

Although Europeans and Japanese are assuming that
breeder reactor programs are their path to energy independence, breeders are not now a commercially demonstrable
proposition, and will not be significant until after the
year 2000. Moreover, there is likely to be a slow rate of
breeder introduction, owing primarily to low stocks of
plutonium. Introducing breeders will also be a slow process
because of the limited risks utility companies are willing
to assume in such a new technology. The Japanese and
Europeans were nevertheless unwilling to accept the
American observation that "breeders breed slowly," and
that their impact will be marginal until at least the next
century.

32

The basic disagreements between the United States and its allies over breeder reactors and enrichment and reprocessing technologies are not, however, about their commercial justification. Rather, they are about their implications for proliferation of nuclear weapons. The allies take considerable offense at implications in American policy that countries taking the breeder route will break their international commitments and pursue a nuclear weapons policy. Acquiring plutonium does not mean a country is contemplating a bomb, even though having plutonium in hand gives it that option. The decision to exploit the option, however, is political, and if a country has made the decision to "go nuclear" in a military sense, there are other avenues in addition to the breeder. What Europeans in particular seem to find irritating is the lack of trust inherent in the American position. They argue that the United States must distinguish between those countries that will handle nuclear technologies responsibly (i.e., the European states), and those that might not.

A final area of dispute between the United States and its allies resides in the question of formulating rules for the nuclear system. Who makes and changes those rules? From the European perspective, the only real answer is a multilateral approach involving not only suppliers but potential recipients as well. This emphasis on multilateralism clashes with basically unilateral measures inherent in recent U.S. nonproliferation legislation. Europeans argue that this action by the United States introduces new rules into the game by national action in contradiction to the U.S. pledge not to indulge in unilateral acts or change existing international agreements. To be sure, U.S. nuclear policy has raised useful questions about nonproliferation. However, the unilateralism of that policy poses serious problems. First, the U.S. action not only regulates relations between the United States and its customers; it also inserts the United States into the bilateral dealings of other states. Clearly, Europeans and Japanese object to such American interference.

A second objection to U.S. policy as defined in new legislation is the uncertainty it creates in the minds of European and Japanese energy planners. In effect, the new U.S. legislation makes potential recipients of American uranium dependent on continuing U.S. approval, which might be forthcoming today; but in 20 years, will a U.S. president also approve shipment of American uranium to Europe? Energy planners are making decisions involving huge investments, and they need more certainty than current U.S. policy provides.

In deciding on its response to U.S. legislation, the European Economic Community considered three strategies.

First was to accept the U.S. legislation, necessitating
renegotiating existing agreements. This course was viewed
as unacceptable and rejected. Acceptance would imply
consent to U.S. unilateralism. In light of such consent,
other countries, e.g., Canada, might follow the U.S.
initiative, thereby beginning a process that could not be
stopped. Moreover, acceptance of U.S. legislation would
contradict the European's pledge not to alter existing
agreements.

The second strategy considered by the EEC was to
reject the U.S. legislation. This strategy was also dis-
carded because most EEC members believed attendant foreign
policy costs too severe. The fact that this option was even
considered by American allies, however, might indicate something about the deterioration in the situation.

The third strategy--adopted by the EEC--was labelled
a "cooperative dialogue." While not accepting the concept
of renegotiation of existing agreements, EEC members decided
to demonstrate a willingness to talk about some of the
issues. In doing so, they are pursuing what they consider
a workable compromise under the circumstances.

On the American side, it was conceded that U.S. policy
probably constitutes a unilateral challenge on nuclear policy.
However, the United States is simultaneously advancing "an
international, multilateral response" to that challenge:
the International Nuclear Fuel Cycle Evaluation, which provides tremendous opportunity for members of the global community
to deal together with problems related to nuclear energy.

The INFCE is addressing three major factors: time,
technology, and institutions. On the first issue, basic
questions are: how much time is available for the development of nuclear energy? What constraints exist globally and
in specific countries? How serious are those constraints?
It is hoped that through the Evaluation's answers there will
emerge a greater appreciation of the time frame within which
nuclear energy can be realized.

In terms of technology, INFCE is looking at a number of
alternatives. It will concentrate on tradeoffs between the
cost of technologies and their implications for nuclear
proliferation. If the cheapest technologies are bypassed
for something more expensive, what gains are achieved in
increasing the difficulties of diverting nuclear materials
from their intended purposes? The Federal Republic of
Germany, for example, is developing a diversion-resistant
chemical enrichment technology. The Evaluation is examining
the extent to which alternatives such as these should be
pursued.

Two types of institutional innovation are being considered during the Evaluation process: first, those assuring
that benefits of nuclear energy are widely available to all

countries even if they do not develop more sensitive parts
of the fuel cycle in national programs, and second, those
for joint control of those technologies that cannot be safe-
guarded effectively on a unilateral basis. Multinational or
regional reprocessing or enrichment facilities would be
included in this second category.

In considering these institutions, participants in the
Evaluation must ensure that whatever institutions are created
are not just "fig leaves" covering national initiatives.
Several barriers to, and high political costs for, abrogation
of agreements regarding these institutions must be created.
Joint management of nuclear projects or having foreign
nationals involved in a country's operation of nuclear power
facilities have been two suggestions for overcoming this
problem. Any such approach is likely to involve sacrifice
of national sovereignty to some degree. Members of the
international community must be convinced that such a
sacrifice is worth the cost.

INFCE provides a two-year period in which to seek agreements which build on the Nuclear Nonproliferation Treaty
(NPT) and the safeguards of the International Atomic Energy
Agency. If too much time is spent on details, however, the
larger political importance of the exercise will be lost from
sight. As a result, there ultimately could be a decline in
international cooperation--the complete antithesis of the
goal that the Evaluation is intended to foster.

If INFCE is successful, the role of nuclear energy will
be clearly enhanced. If success is not achieved, however,
existing problems will be exacerbated and new ones created.
Both the United States and its allies must hope that it will
contribute to amelioration of their differences.

Underlying the overall debate between the United States
and Western Europe and Japan is the question of adequacy of
current American nuclear policy. While the United States has
raised important issues, it has also created turmoil within
the nuclear industry and tensions not only with its allies,
but also with other countries such as Yugoslavia and India.
One expert clearly outlined the consequences of this turmoil
and tension when he said:

A period of turmoil and tension can be productive,
but continuing uncertainty can be disastrous. If
U.S. electric utilities continue to forego ordering
nuclear power plants because of licensing delays
and uncertainty concerning the disposition of
radioactive waste, the United States could lose its
option to continue to expand rapidly the domestic
use of nuclear energy in the 1990s when such
expansion may be needed. If the terms and conditions imposed on U.S. nuclear exports are not
viewed by importing countries as acceptable and

stable, U.S. manufacturers, and the government itself
in the case of uranium enrichment, could find themselves unable to sell in the export market for
nuclear fuel, equipment and services at a time
when such exports are important for employment
and balance of payments. If other countries dis-
connect from their previous longstanding nuclear
energy cooperation with the United States, the U.S.
Government could lose much of its leverage over the
weapon proliferation problem. Finally, in the long-
run, if other nations strive to achieve self-
sufficiency and leave the framework for international nuclear cooperation which the United States
and others have struggled long and hard to evolve,
then the risks of weapons proliferation will be
greatly increased.

U.S. nuclear policy, then, can be the key to the evolution
of the nuclear power and nonproliferation regime that emerges
in the next two decades. It will also have a major impact
in structuring American political relations with the rest of
the world.

Some problems inherent in current U.S. nuclear policy
stem from the fact that the policy appears to break sharply
with the past. For example, at the Second Geneva Conference
in 1958, there was an argument whether nuclear testing should
be stopped. Japan argued that in addition to weapons testing,
peaceful nuclear explosions should also be halted. The United
States opposed this, insisting that PNEs not be included in
any ban. It was argued that, ultimately, that policy provided
the opportunity for India to explode its nuclear device in
1974, a development the United States finds unacceptable.
Similarly, in terms of the use of plutonium, the United
States formerly insisted that Japan use plutonium for light-
water reactors to save enrichment capacity. The Carter
Administration now opposes such use. While each U.S.
administration obviously has the right to determine its own
policy, these recent reversals of long-standing U.S. policy
have provided serious problems for American allies.

Some U.S. spokesmen argue, however, that there has been
no real discontinuity in American policy and that U.S. goals
remain the same. The situation in the mid-1970s changed
drastically; the year 1974 represented the end of what one
participant labelled "the happy era." In the wake of that
year's Indian PNE, the question had to be faced, whether a
distinction between commercial and military uses of the atom
could still be drawn. In Canada, which had supplied India
with its nuclear fuel, public opinion demanded tighter
measures to control nuclear exports, a demand that eventually
led to the ban on Canadian exports even to Europe. In the

United States, the Congress of 1974-75 appeared ready to
pass even more stringent measures than it did recently.

A second major development in 1974 that altered the
energy environment and generated calls for a new U.S. policy
was the Arab oil embargo. In its aftermath, projections of
nuclear demand were grossly inflated. They were not at all
based on estimated elasticity of supply and demand. As a
result, many 1974 projections generated a "full speed ahead"
planning syndrome in many countries with little regard for
other consequences. One expert called these projections "an
international myth to justify the premature spread of sensitive reprocessing facilities before they had any economic
justification." If security was considered at all, it was
relegated to tertiary importance. The international system
had not had time to develop safeguards, while the need for
control of the security problems created by the new situation
had intensified.

By the time President Carter took office, it was argued,
many countries were basing their increased emphasis on use of
plutonium on assumptions that were not necessarily correct.
At the same time, there was widespread public disagreement
on the utility of nuclear power as a result of the Indian
PNE. The Carter Administration believed that this new
situation called for a policy with different emphasis.

In developing its nuclear policy, the basic problem
confronting the Carter Administration is the tradeoff that
must be made between requirements imposed by the need for
nuclear power and risks of a potential proliferation of
nuclear weapons. The same materials are used in both bombs
and peaceful reactors. How can one ensure that nuclear power
will take its place as an important source of the world's
energy and that the number of countries with nuclear weapons
will remain limited?

Of course, the current administration is not the first
to grapple with this problem. The dilemma has been apparent
since 1945 when the Acheson-Lilienthal Report was issued.
Since that time a series of efforts to solve it--beginning
with the Baruch Plan itself--have been undertaken. The
United States tried early on to keep the technology to itself,
but by 1954, this was seen to be impossible. A change in
policy was effected, culminating in the Atoms for Peace
Program which sought to keep a distance between military
and commercial uses of nuclear technology through a system
of international safeguards.

Today, the basic philosophy of the Atoms for Peace
Program remains essentially correct. There is no long-term
future in a denial philosophy. Problems have arisen, however,
because it is becoming increasingly difficult to distinguish
between the peaceful and the military atom. One must draw
a distinction between a situation in which a government
decides to initiate a military nuclear program from scratch

and one in which it is brought to the stage of possibly producing nuclear weapons through evolution of its peaceful nuclear energy program. A large-scale transfer of nuclear materials and nuclear technology contributes to this latter scenario, and it is to prevent such an occurrence that the policy of the United States and other Western countries must be designed. It was asserted that the nuclear technologies that countries such as France and the Federal Republic of Germany have tried to sell in the Third World--enrichment and reprocessing technologies--are the most dangerous parts of the fuel cycle for weapons proliferation.

As mentioned earlier, however, the decision to procure nuclear weapons is a political decision resting more on will and desire than on access to technology. It was estimated, for example, that approximately 35 countries, if they choose to, can make nuclear weapons outside the nuclear fuel cycle. They can develop weapons within four or five years of the decision to go ahead with a program and at a cost of about $50-100 million.

Since the decision to "go nuclear" in a military sense depends less on existence of nuclear power plants than on political considerations, measures to prevent proliferation must also be political. "Costs" for going the nuclear weapons route must be made too high in political terms to be perceived as worth the risk. Positive incentives must also be provided to encourage countries to limit nuclear programs to peaceful uses.

This is not to argue that transfer and use of nuclear technology should not be carefully controlled. On the contrary, there is a strong need for extraordinary care in managing nuclear technology and nuclear materials, not only because of proliferation, but also to prevent accidents. Diversion of nuclear materials to military purposes or an accident causing massive damage anywhere could have negative consequences everywhere, especially for the nuclear power industry. The adverse impact of the Indian PNE on perceptions of the utility of nuclear power is only one recent example of the problem.

Prospects of a world in which nuclear weapons are proliferating rapidly are unsettling at best. At the same time there is little doubt that reliance on nuclear energy, on breeder reactors and reprocessing technologies will increase in the world at large. How can benefits of nuclear power be best exploited without increasing the dangers of weapons proliferations?

In short, where do we go from here?

One expert proposed a system of commercial enrichment and reprocessing on an international cooperative basis. Such a system would entail mutual fuel dependence in which industrial countries provide enriched uranium for light-water reactors to less developed countries which in turn

would contribute plutonium to fast breeder reactors of more
advanced countries. This interdependence would not develop
overnight, of course, but take decades to evolve fully.
It was argued that such a system would have several benefits:
first, it can be justified on commercial grounds; second,
less developed countries would not be denied access to
nuclear power and, where feasible, multinational facilities
might even be constructed in regions of adequate markets;
third, and perhaps most important, proliferation risks of
such a system should be acceptable to all, including the
United States (unless it decides a plutonium economy anywhere
would be too risky).

Of course, this is only one of several alternatives.
Whatever system does evolve, the United States cannot afford
to stand in isolation. It has played a leading role since
the beginning of the nuclear era in emphasizing nuclear
nonproliferation and in promoting nuclear power. There are
compelling reasons for continued U.S. leadership in both
the nuclear industry and in the drive to halt the spread of
nuclear weapons. Delay and indecision on Washington's part
will benefit no one. The United States must assert strong,
consistent and pragmatic leadership in a multilateral effort
to take advantage of the immense benefits of nuclear energy
without increasing its already substantial risks.

*   *   *   *   *   *   *

SUMMARY STATEMENT BY THE
HONORABLE MIKE McCORMACK

The nuclear energy panel at the Fourth Quadrangular
Conference provided considerable insight with respect to the
complex international political environment dealing with
nuclear energy. While the danger of nuclear weapons proliferation increases (with every new addition to the nuclear
energy club), the United States lacks a realistic international nuclear energy policy which provides leadership to
the Western and Third World nations. There is a requirement
for the United States to provide international leadership in
developing a consensus with regard to the safe employment
of nuclear energy.

The need for well-ordered development of nuclear energy
has been intensified by the increased price of petroleum
and the threat of a new embargo by the Organization of
Petroleum Exporting Countries. Several countries face
more intense pressure than the United States to free themselves of a dependence on OPEC petroleum. Japan and the
nations of Western Europe have turned to emphasis on the
development of the breeder reactor as one alternative.
The breeder, which produces more fuel than it consumes,

39

offers the prospect of energy independence. My own view is
that the breeder can provide secure energy supplies worldwide with a minimal threat of nuclear weapons proliferation.
If appropriate controls are exercised over the handling
and reprocessing of nuclear fuels, including the manufacture of new fuel elements from reprocessed fuel, the
breeder cycle will not cause a significant or unique threat
of nuclear weapons proliferation.

The recently proposed administration policy designed
to limit the threat of proliferation hinges on the deferral
of nuclear fuel reprocessing and commercialization of the
breeder reactor. As much as I appreciate the President's
concern for nuclear weapons proliferation, his proposed
policy is based on several false assumptions. The first
is that a research program involving the breeder reactor
would, in itself, increase the potential for nuclear proliferation by increasing the availability of weapons grade
plutonium. In the United States alone, about 500 times more
plutonium will be produced each year from conventional
nuclear power plants in the late 1980s and early 1990s
than from a breeder program during the same time. No
weapon has ever been constructed from nuclear material
originating in a commercial power plant. It is much cheaper
and easier to construct a weapon using facilities built for
this purpose than to try to divert nuclear fuel. Several
dozen nations now have the capacity to produce nuclear
weapons without having any nuclear power plants at all.

The second misleading assumption is that U.S. post-
ponement of breeder development will induce other nations
to do likewise. It is difficult to believe that nations
which have committed themselves to breeder development as
an answer to heavy dependence on foreign petroleum supplies
would follow us in postponing research and development on
this alternative source of energy. Japan and most Western
European nations do not possess the abundant supply of
fossil fuels available to the United States. Moreover,
the U.S. supply of uranium is an unknown, and it may be
that there will never be a sufficient supply of U.S.
domestic uranium to export to nations depending on
standard nuclear power plants. Leadership by the United
States requires that we avoid asking our allies to discard
one of their chief hopes for energy independence.

Another false assumption regarding the breeder research
program is that the present system is significantly more
susceptible to diversion of fuel to weapons than some
alternative breeder system which is technologically
feasible. This is not true. The complexity and difficulty
of extracting plutonium from fuel can be made even more
difficult than it is now. Together with appropriate
institutional safeguards, such physical difficulties
reduce the probability of proliferation to insignificant

levels. Of the presently known proposed breeder alternative
technologies, none is as attractive as the Liquid Metal
Fast Breeder from an economic or engineering standpoint,
and none significantly better from a proliferation standpoint.

The Federal government must make decisions in this
as well as other areas by employing the best available
scientific, engineering, and economic data. This course
will lead us to establish a realistic energy policy which
provides us with necessary energy at the lowest possible
cost.

PANEL IV

FACING THE ECONOMICS OF THE ENERGY PROBLEM

Economic issues surrounding international energy
were discussed by a panel chaired by William D.
Rogers of Washington's law firm of Arnold and
Porter, and former Under Secretary of State for
Economic Affairs. Members of the panel included
George Melloan of the *Wall Street Journal*;
Thierry de Montbrial, Director of Centre d'Analyse
et de Previsions, French Ministry of Foreign
Affairs; Alirio Parra, a director of Petroleos
de Venezuela, S.A.; and Edwin H. Yeo, Chairman
of the Asset and Liability Management Committee,
First National Bank of Chicago, and former Under
Secretary of the Treasury for Monetary Affairs.

FACING THE ECONOMICS OF THE ENERGY PROBLEM

The impact of the sudden fourfold increase in imported-
petroleum prices in 1973-74 shook the industrialized and
Third World countries as no other economic development has
since World War II. While major industrialized nations
suddenly were paying tens of billions of added dollars
for imports, poorer Third World nations were required to
pay, proportionately, even more burdensome sums for
petroleum supplies, as well as to cope with energy-
related shrinkage of industrialized markets for their
products. The resulting problems of economic stability
and continued economic growth have been massive. Clearly,
they must be addressed at every level--private firms,
individual governments, and international organizations--
if they are to be successfully met. Although the international financial system has shown unsuspected resilience
and flexibility since 1973, there is no guarantee that
further substantial shocks can be accommodated in the future.
    In the face of already massive petrocurrency flows,
the viability of the international financial system itself
is in question. According to popular theory, the 1974
recession, and a great deal of the world's inflation over
the past few years, should be attributed solely to the steep
jump in petroleum prices engineered by the Organization of
Petroleum Exporting Countires (OPEC). There have been
other factors, however, that have contributed to the
problems of the world economy.
    Poor management of industrial economies, long before
the OPEC price rise, put these countries into weak positions.
For example, in the United States, the consumer price
index had been increasing rapidly since the mid-1960s. In
1971, price controls were instituted to halt the trend, but
at the same time, the domestic money supply was being
increased at a rate above 7 percent. It was not surprising,
therefore, that with removal of controls, prices climbed
rapidly in the 1973-74 period. This would have happened
with or without the OPEC cartel's successful maneuver.
    Similar mismanagement of the domestic economies of
several other countries in the Organisation for Economic
Co-operation and Development (OECD) could partially explain
price rises in these nations as well. Between 1973 and
1975, for example, the money supply in the United Kingdom
increased by over 14 percent. With industrial production
dropping by almost 4 percent during the same period, the
corresponding 20 percent increase in consumer prices can
be explained with little reference to OPEC pricing policies.

In France, during the same period, consumer prices rose
almost 13 percent, partially in response to an increase
of over 11 percent in the money supply.
These expansive monetary policies were pursued to
stimulate sluggish economies through demand increases.
In Latin America and Europe, governments had to contend
with social disorder, which made stimulation even more
essential as a method of creating jobs. According to
one expert:

> It is difficult to separate cause and effect in
> terms of a quadrupling of oil prices. We were
> on a course that resulted in very substantial
> rates of price increase in most of the principal
> economies of the world. Disequilibrium as measured
> by a structural imbalance in current accounts had
> appeared or was clearly incipient in a number of
> countries that had enjoyed relative stability in
> the sixties. Italy is a case in point. A
> marvelously managed economy in the sixties,
> running a structural current account surplus,
> it moved to a so-called "stimulus" policy in the
> first part of the seventies, well before the
> quadrupling in the price of oil. So the way that
> I would look at it is that a quadrupling of oil
> prices fell on an international monetary system
> that was already under serious strain because of
> disequilibrium in key sectors.

While the increasing cost of energy imports has clearly
contributed to the current-account deficits of industrial
nations, several countries already had current-account
deficits prior to OPEC price increases. In 1972, for example,
only Germany among the major industrialized nations exported
more than it imported. Thus, while OPEC price increases are
supposed, by many, to have initiated the problems of
inflation, unemployment, and current-account deficits for
oil-importing nations, others have argued that this is an
unwarranted accusation. Further, they argue, price increases
had positive effects because they helped to bring under
control the world's runaway use of petroleum before a severe
shortage could develop. This was the reasoning of one
panelist, who summarized the situation as follows:

> Prior to 1973, we were all aware of the inflationary
> pressures in the international economy. Monetary
> policies in many countries were inflationary;
> bottlenecks were appearing everywhere. There was,
> at the same time, a unique coincidence of high
> economic activity in all the OECD area. And
> together with this, there was runaway consumption

of energy.... Consequently, these pressures led
to an increase in prices. The increase resulted
in part because of the demand pressures and in
part because the producing countries were becoming
progressively disillusioned with the deteriorating
terms of trade that were appearing in the world
economy. Many would have preferred not to have
seen a sudden, dramatic increase in prices as we
witnessed in 1973, but rather, a gradually increasing trend over a longer time span. Many in
the producing countries became disillusioned,
seeing oil prices, in real terms, declining. The
few voices heard were like voices crying out in
the wilderness, trying to point to the need for
higher prices, for conservation, and for funds
that could go into future energy supplies.

The United States inevitably plays a key role in helping
to control international economic disequilibrium, because
of the country's large size and its contribution to international trade. Since the U.S. dollar is the major instrument in conducting international trade, the United States,
by establishing reasonable dollar stability, can help ease
current and future financial strains. A first necessary
step for the United States is to constrain the long-term
growth rates of energy imports. Success in doing so would
have substantial benefits, both in easing world demand for
petroleum and in lessening pressure on the dollar in
international currency exchanges.
Some economists and monetary experts argue that the
decline in the value of the dollar is due less to energy
spending, and more to continued poor macroeconomic management, as evidenced in the U.S. high-budget deficits and
rapid monetary growth. Recent months have witnessed
increases in U.S. consumer prices as high as an annual
rate of 10 percent. This rise has accompanied a rapid
increase in the money supply. There is wide agreement
that both domestic inflation and continued large imports
of petroleum contribute to the dollar's slide. Both need
urgent attention in the months and years ahead.
Many analysts predicted that the first victim of the
OPEC price increase would be the international monetary
system. The major oil-price increase had enabled many of
the producing countries, especially Saudi Arabia, to build
up large surpluses of foreign currencies. The relatively
underdeveloped character of most of these producer-nation
economies would not permit rapid investment domestically
to match the pace of petroleum revenues. Although producer
nations found themselves in an extremely liquid position
with regard to international currencies, long-term investment

47

in foreign markets was a difficult process for several
reasons, among them the continuing rapid decline in the
value of foreign currencies. OPEC investors did not
want to acquire assets producing returns that were being
severely devalued by inflation. Accordingly, they
resorted to short-term accounts in banks of industrial
nations. But these assets could hardly provide a hedge
against inflation. In addition, the banks could not
indefinitely accept short-term money from OPEC investors
and, on the other side of the balance sheet, make long-
term loans.

The producer nations with foreign-currency surpluses
also have a continuing problem of coping with distrust
in the industrial nations about their investments in
real estate and public corporations, for example. Since
they have met with some hostility in Europe, they are
making every effort to avoid negative publicity in their
U.S. investments, where the press, Congress, and regulatory
agencies are monitoring their financial moves.

The international financial system faced, and still
faces, an uncertain future, too, because many oil-consuming
nations have been forced to finance imports by borrowing
extensively in international financial markets. Such
was the case in Italy and in the United Kingdom. The
borrowing needs of the less-developed countries (LDCs) have
been particularly acute. Yet, there was reluctance on the
part of OPEC governments to make their excess reserves
available for such borrowing. Since the initial oil-price
increase, some progress has been made in convincing OPEC
money managers to make funds available to various international lending mechanisms. But substantial reliance has
had to be placed on commercial conduits to recycle these
funds. Most experts see little cause for near-term alarm in
these arrangements, unless OPEC funds begin shifting as a
result of political instability. At present, about two-thirds
of OPEC wealth is invested in short-term bank and treasury
instruments. Of the remaining assets that are placed in
long-term investments, about half is found in U.S. securities.
These recycling responses have demonstrated the flexibility
of the international financial system, but there is real
question whether they are stretched to their limits at
this point and unable to flex further to meet additional
demands.

Responses of various consuming nations to the "petrodollar crisis" have varied considerably. The United Kingdom
has come to depend on its own oil production in the North
Sea; the United States has become a favored place for OPEC
investments in one form or another; several European
countries have substantially reduced oil imports by using
petroleum more efficiently while also accepting petrodollar-
investment funds; the less-developed nations have managed

to survive through a combination of adjustments, particularly
through commercial-bank borrowing.

As previously noted, a large portion of the petrocurrencies is returned to their countries of origin by
virtue of capital transfers of different types. In addition,
a major share of the funds is recycled through the oil
producers' purchases of capital goods and consumer products.
Capital goods are particularly attractive to most OPEC
nations because their economies, for the most part, are
narrowly based on oil exports, and diversification will be
necessary for long-run growth once petroleum reserves are
exhausted. Funds are also recycled through purchases of
military equipment from the industrial nations. The Iranian
government, for example, was noted for this employment of
foreign-currency surpluses. Other nations, including Saudi
Arabia, with its massive overseas purchasing power, have
opted for similar purchases with their petrocurrencies.

Fortunately for major industrial nations, the floating
exchange rate system, in effect since early 1973, has
provided some of the flexibility needed to adjust to the
enormous strains of the oil price increases. Exchange
rates between oil-producing and consuming nations changed,
to held accommodate surpluses, and exchange rates between
consuming nations also shifted, in order to redistribute
cpital flows directed heavily against a few major industrial
powers. Of course, floating rates are not a complete answer
to persistent payment imbalances, for a considerable proportion of world trade is relatively insensitive to price
changes--oil, in particular, has proven to be less price
elastic than many had hoped.

For the most part, the urge to adopt "beggar-thy-
neighbor" policies was restrained during the worst of the
crisis. One exception was in Italy, where the government
imposed import restrictions almost immediately after OPEC
price rises as a means of coping with the Italian balance-of-
payments problems. The restrictions impacted on products
other than petroleum, such that the trade deficit was thrust
onto countries that regularly exported these other goods to
the Italian market. Elsewhere, too, there was some
succumbing to protectionist pressures. Generally, however,
governments of industrial nations relied primarily on
forced conservation measures and higher energy prices to
reduce demand for petroleum imports.

Besides inducing increased emphasis on conservation,
OPEC price rises have stimulated energy production outside
OPEC boundaries. Since 1973, petroleum in commercial
quantities has been discovered in ten additional nations,
bringing to 23 the number of non-OPEC countries with proven
reserves. Additionally, serious efforts at exploring for
various types of energy have been launched, since 1973, in
35 non-OPEC countries. OPEC officials note that all of these
reactions to price rises will be beneficial to the world's
economic health.

These officials are also anxious to point out that OPEC members have not engaged in speculative practices. This contention, of course, might be challenged by some who would define 1973 price increases as speculative moves. Be that as it may, the official OPEC position has been that members have acted responsibly by providing loans to oil-consuming nations, especially the LDCs, to aid them with their balance of payments.

If any blame is assigned in connection with price increases, OPEC nations point to pre-1973 oil-consumption levels of the industrial world. Wasteful consuming practices and overheated economies of several of these nations generated record levels of demand. Thus, they claim, blame for sudden price increases should be placed on governments that tolerated waste and employed the stimulative monetary and fiscal policies that eventually overheated these economies.

Energy pricing is an important issue on two accounts: first, appropriate pricing is essential for efficient allocation of resources; second, questions of equity, income distribution, and rests--all price related--are important from a political standpoint. A retrospective glimpse at petroleum pricing over the last few decades helps one to understand today's market.

Until World War II, pricing of the world's petroleum supply was responsive to controls established by the regulatory agencies of the U.S. producing states and to the gradually expanding discovery of oil in the Middle East. By the 1930s, fear of a petroleum shortage had disappeared, and reduction in overall economic activity during the Great Depression had significantly reduced demand. In response, the U.S. government imposed restrictions on importation of foreign crude to help keep domestic prices higher, and the Texas Railroad Commission and other state regulatory agencies in the United States imposed restrictions on domestic production.

World War II expanded demand for oil needed to fuel the war machines of the industrial nations. In turn, increased demand encouraged the search for new sources of supply, so that after the war a new industry structure developed. Until the end of the conflict, most of the world's petroleum was produced by a few firms. However, after the war, as world economic growth increased demand, new firms joined the search for, and production of, crude petroleum, as well as the refining of various products. Not only did private firms expand into the international oil business, but state-owned firms also entered the competition for world markets, in response to an increasing surge of nationalism and a desire by various governments to earn foreign exchange for postwar development activities.

As a result of this additional market entry and a subsequent expansion of supply, world petroleum prices fell in

the postwar period until the end of the 1960s. In the U.S.
market, import restrictions were again imposed in 1959 to
bolster prices for American producers who were receiving
increasing competition from the Middle East. Under the
rules of this mandatory oil-import program, import shares
for various firms were set at then-prevailing proportions.
The result was a U.S. domestic petroleum price higher than
the world price, and increased returns for domestic producers.
However, falling world prices and the loss of a large share
of the U.S. market experienced by developing oil exporters
of the world led to a response by these exporters--the
formation of OPEC in 1960. In September of that year, Iran,
Iraq, Kuwait, Saudi Arabia, and Venezuela formed OPEC as a
means of achieving bargaining power with multinational oil
companies.

By the early 1970s, control of the pricing of internationally traded oil clearly had fallen into the hands of the
OPEC nations. What would otherwise have been a strategically
difficult task of trying to form a collusive pricing arrangement among oil-exporting nations was made feasible by rising
demands and the concentration of a large portion of producing
capacity in the hands of a single nation, Saudi Arabia.
With output capacity exceeding 8 million barrels a day, but
with limited financial needs, the willingness of the Saudi
government to accept production cutbacks made possible the
establishing of a significantly higher cartel price.

Since the 1973 price increase, normal market incentives
have produced additional competition for the cartel. At this
writing, expanded production in non-OPEC nations as well as
price-induced conservation have forced the cartel to halt,
at least temporarily, nominal price increases. Since OPEC
prices are quoted in U.S. dollars, whose value has been
plunging, the real price of the cartel's production actually
declined in response to the so-called petroleum glut. This
has not stopped several OPEC nations from threatening
additional price increases in the near future. Saudi Arabia,
however, has thus far opted to maintain the status quo and
to permit real prices to decrease. Of course, this may not
remain the Saudi position; the lack of inflation-proof
investments in foreign markets, and limited capacity for
capital absorption at home, could convince the Saudi government to reduce production in order to hold reserves for the
future.

The efficiency of resource allocation through the price
mechanism has been partially ignored by present U.S. policy.
The fact that a full market-price system is not permitted to
function means that the most efficient transitions among
oil and gas and other fuels will not be made. Explicit
costs that will likely be borne for interfering with the
market include premature abandonment of oil- and gas-related

51

capital goods; the requirement to employ partially developed
technologies for alternative sources of energy, and the
failure to provide sufficient lead times to achieve levels
of conservation associated with the consumption of higher
cost fuels.

Price controls on U.S. domestic production have had
the unintended effect of bolstering the effectiveness of
the OPEC cartel. Since the power of the cartel is dependent
upon low elasticity of demand for its product, the most
effective counter to the cartel would be to encourage supply
competition, which would include maximum U.S. domestic
production of oil and gas. The real price decrease we
witnessed during 1977 and 1978 might have been larger had
the United States more fully stimulated domestic production
since 1973. While formal price-control mechanisms are
likely to remain in place for the near future, their effectiveness may be purposely decreased, according to some analysts.
One expert observed:

> I think that we are going to cope gradually with
> the politics of the energy situation and we will
> decontrol energy prices in fact, if not in appearance.
> We are working toward that, obviously, with the
> energy package. I think that we have to get there
> eventually because it is important to return to
> some market system for deciding what kind of energy
> we develop, and how we get the optimum price in
> the energy that we use in our economy. The only
> way that this can be done is through letting the
> market make these choices of priorities.

Part of the problem in establishing an effective U.S.
policy response to cartel price increases has been finding
the answer to the question, who should receive the "rents"
associated with the new world price? In this usage, rents are
revenues received in excess of costs borne by producers in
finding and producing petroleum supplies. Quite obviously,
rents from OPEC production are accrued by governments concerned. In the United States, rents for older oil wells are
held down by price controls. Politically, this is a popular
policy, but it makes little sense when additional production
from wells of all vintages needs to be encouraged and when
conservation should be fostered among consumers. The goal
of avoiding large income redistributions--the reason for the
price controls--could be better accomplished through policies
designed to tax producers and distribute proceeds in some
manner that does not encourage additional consumption.

In trying to describe the correct or best price for
petroleum in international or domestic markets, one has to
recognize that history provides little guidance. Thus, while
a free, competitive market is the optimal mechanism for

settling prices, from an economist's viewpoint, a completely
free market in oil has never existed. It has either been
a market of few private producers or of governmental
controls. However, in looking to the future, the government should pursue the goal of achieving a market as free
from controls as is feasible.

In summary, it will be necessary for governments of
oil-consuming industrial nations to establish policies that
provide for replacement-cost pricing of energy resources and
to develop an international monetary system that is flexible
enough to handle massive petrocurrency flows. Despite the
temporary reduction in the real price of international oil,
new, real increases are probable in the not-too-distant
future. In the absence of sensible forehanded policies,
the industrial nations might well again find themselves with
their standards of living sliding downward, with high rates
of simultaneous inflation and unemployment, and with a
staggering international monetary system. The triggering of
major price increases could come early in the 1980s, when
the world's demand for oil presses against the likely limits
of world production. Given the lead times required to bring
forth alternative sources of energy in significant amounts,
and given the adjustments necessary to reduce energy consumption, effective policies have to be instituted now to
avoid the problems cited.

The nations of the non-Communist world look to the United
States to put its energy house in order for three reasons:
first, and most importantly, to ease the demand pressure on
the international oil market and reduce the burden for other
nations; second, to reduce its dependence on OPEC oil in
order to free American foreign policy from narrow outside
influence; and third, to stanch the flow of inflationary,
turbulence-producing dollars throughout the international
economic system. This latter concern is part of a yet broader
set of concerns, namely, the hope that the United States, as
the still-acknowledged leader of the industrialized democracies, will adopt a comprehensive international economic
policy--embracing energy policy as one element--in order to
cope with international monetary and economic turbulence.

*   *   *   *   *   *   *

SUMMARY STATEMENT BY THE
HONORABLE WILLIAM D. ROGERS

It is unlikely that anyone who participated in the panel
discussion agreed entirely with everyone else; the subject
is too exciting to allow dull unanimity. By the same token,
no summary of the discussion could be expected to command
the unqualified support, in all particulars, of either the

53

individual panel members or the chairman. Nevertheless,
having, as that chairman, profited so considerably by the
discussion and the contributions of the principal participants, I feel that this view captures the essential
character of the discussion and, by the same token, is a
contribution to public consideration of the issue.

My own views, considerably sharpened and informed by
the panel meeting, are as follows:

> (1) The quadrupling of energy prices was a major,
> but by no means the exclusive, cause of recent
> world inflation and growth-rate trends.

> (2) Although a national energy policy is very
> much in our interests and the world's, it will
> not come anywhere near making us independent of
> OPEC oil.

> (3) We must therefore adjust to the new inter-
> national economic realities in this new era of
> expensive energy and divergent inflation and
> growth rates by adopting a comprehensive inter-
> national economic policy.

> (4) This policy must concentrate not only on U.S.
> energy-import restraints, but also on the harmoni-
> zation of our own growth and inflation rates with
> those of our major trading partners; otherwise,
> no solution to the present payments imbalance is
> in sight.

PANEL V

THE FUTURE OF COAL: AN AMERICAN CASE STUDY

The future prospects for coal supply and demand
were addressed by a panel of four representatives
of various groups interested in coal's role in
meeting world, and particularly U.S., energy
requirements. Albert Gore, formerly United
States Senator from Tennessee and now Chairman of
the Board of Island Creek Coal Company, was in
the chair. Other panelists were Ralph Bailey,
President, Continental Oil Co., and former Chairman
of Consolidation Coal; Larry Moss, Chairman of the
Environmental Caucus of the Center's National Coal
Policy Project and a former president of the Sierra
Club; and Joseph A. Yablonski, Jr., partner in the
labor law firm of Yablonski, Both and Edelman.

THE FUTURE OF COAL: AN AMERICAN CASE STUDY

The energy market in recent years has provided incentives
for the world to reverse the long-standing trend to less coal
consumption. Coal has also benefited from the desire of
many nations to reduce reliance on petroleum imports from
the Organization of Petroleum Exporting Countries (OPEC).
The announced policy of the Carter Administration is to
require conversion of central power stations in the United
States from oil and natural gas to coal. The President's
goal calls for consumption of 1.2 billion tons per annum by
the year 1985, roughly double the current figure. The panel
was agreed that such a goal is feasible--with a number of
important "ifs," that is.

Potentials for employing coal as a substitute for imported petroleum are excellent--at least in theory. Coal
comprises 90 percent of U.S. energy reserves, yet it provides only 19 percent of present energy supplies. Estimates
of U.S. reserves run as high as four trillion tons, at
current consumption rates, a 600-year supply. The chief
barrier in expanding coal markets exists on the demand side,
since physical capacity to expand production appears to be
no problem at this time, assuming, that is, improved regulatory
policies on the part of government.

Coal was the backbone of America's industrial growth
in past decades. At the turn of the century, coal supplied
over 90 percent of the nation's energy demand. Markets for
coal grew steadily through to the mid-1940s, and in 1947
the industry produced 630 million tons, a production record
that stood until recently.

While 1947 marked a record for coal production, it also
marked the beginning of a 15-year decline in coal consumption.
Oil and natural gas, fuels readily available and cleaner
burning, largely displaced coal in both railroad and home
heating markets, which were principal markets for coal up
until that time. Despite greatly increased energy demand
by the nation, coal steadily lost ground, hitting a low
point of approximately 400 million tons in 1961.

In the early 1960s, two things happened that provided
new growth for the industry. Demand for coal-generated
electricity surged upward and the nuclear power industry,
which was projected to meet much of that demand, expanded
at a far lower rate than originally predicted. Consequently,
the coal industry found new life and achieved a 50 percent
increase in production between 1960 and 1970.

Production leveled off in the early 1970s, however,
and the coal industry produced only 600 million tons in

1974, compared with 630 million tons almost 30 years
earlier. Then the OPEC oil embargo of late 1973 and early
1974 produced a new spurt in demand for coal, produced by
the cost picture for alternative fuels, namely oil and
natural gas, having changed dramatically. The coal
industry responded with aggressive expansion and set a
new production record of 650 million tons in 1974.

For its part, the coal industry sees no physical
problem in expanding capability to meet national energy
needs. On the demand side, there already appears to be
some voluntary shift to coal by U.S. industry. The electric
power industry now has plans to have 250 new coal burning
power plants on line by 1985. Of course, the shift to
coal by utilities is actually a reversal of the earlier
trend when coal was replaced by cheaper and cleaner burning
petroleum derivatives.

An additional area in which coal can make a contribution to U.S. economic well-being is international trade.
In 1974, exports were 54 million tons, about 8 percent of
production, yielding about $2.7 billion. There is little
expectation, however, that exports of coal will expand
greatly in the near future, for most coal exported is for
a metallurgical use and depends on demand in the steel
industry. Steam coal does not promise to be a major export
item since large deposits are found in several other nations
including the United Kingdom, Canada, Australia, Venezuela,
and the Soviet Union.

The sudden need for coal in the United States as a
substitute for imported petroleum occurred at about the same
time that public attention became focused on improving the
quality of the nation's environment. Consequently, the full
incremental cost of a shift to coal has included the cost
of reducing the levels of pollution in the environment
which increased coal use brings. This component of the
cost is really an exchange of an externalized cost of air
and water pollution for a cost internalized in production
of electricity and passed on to consumers.

Strict reclamation laws, too, may have serious impacts
on coal production and demand. Large parcels of coal
reserves have been effectively eliminated from mining in
the foreseeable future because cost of reclamation would
make such coal prohibitively expensive. Moreover, while
there are large areas in the United States where present
reclamation technology can be usefully employed, there are
some areas, particularly those with inadequate rainfall,
where reclamation is not now feasible even at high cost.

In order to maintain the integrity of the environment,
it may be necessary to limit mining to places where reclaimability is well proven or where agricultural or other land
use productivities are low. "Reclaimability" should include
consideration of underground aquifer quality, future

surface transportation requirements, and the future workability of the sites for recovery of other resources. The
goal of reclamation in arid areas should be the establishing
of self-regenerating natural growths.

Another major uncertainty in the coal industry is the
future policy of the federal government about leasing its
coal lands in the West. Presently, there is no clear
federal policy on this issue, so the courts have ordered
all leasing to cease as a result of environmentalist law
suits. Since the economic feasibility of mining many coal
lands already leased is dependent upon lease acquisition of
adjacent federal lands, further potentially productive
lands sit idle today.

The main environmental obstacles to burning coal are
the Clean Air Act and its 1977 Amendments which require
removal of air pollutants generated by combustion processes
in order to limit discharge of pollutants into the atmosphere. These requirements are being interpreted to mean
that _all_ new coal burning facilities must install expensive
and, many argue, unperfected devices known as scrubbers.
This sweeping interpretation raises questions of cost-
effectiveness. Are we getting the most pollution abatement for each dollar spent? Can we afford to incorporate
relatively high pollution abatement cost in the price of
products that must compete in the world marketplace?

With passage of the Clean Air Amendments, it has
become clear that scrubbers will have to be installed on
all new plants, for they represent the required "best
available" technology at this time. Scrubbers are not,
however, technologically refined and have a tendency to
break down at frequent intervals, which requires that plants
be equipped with backup scrubbers or cease operations if
the primary device fails.

So far no one really knows the full impact of the air
pollution regulations. One major unknown is whether
industrial growth will occur in areas which are now unable
to attain mandated air quality standards or in areas which
are labeled pristine, and thus not allowed to undergo
significant deterioration.

Another unknown is the impact on the environment should
pollutants from increased coal usage reach levels well
beyond those with which we have had experience. One
particularly urgent question, for example, is the impact
of increased amounts of sulfur and nitrogen oxides on human
health and plant production. Large increases in levels of
pollutants will result from the shift to coal encouraged by
the U.S. National Energy Plan, and available technology can
do little to prevent such pollution. Yet the new requirements will not help but will certainly hamper production of
low sulfur western coal, for the Clean Air Amendments are
now interpreted to require scrubbers regardless of the

sulfur content of coal. The economics flowing from this
interpretation favor eastern or midwestern coal reserves.
While these deposits have higher sulfur content, this disadvantage has been mitigated by requirement that all new
plants have scrubbers; it will no longer be profitable for
utilities in order to reduce capital investments in pollution
control equipment. In addition, accuracy of the data upon
which current air quality standards are based is constantly
being challenged, making it difficult to regulate permissible
levels of suspected pollutants. These and other uncertainties
preclude any hard "scientific" forecast of whether we can
"afford" to implement our air pollution laws.

One unresolved issue discussed by the experts is how
best to encourage coal as a substitute for oil and gas.
Congress is currently debating legislation that would man-
date conversion of all existing oil and gas fired boilers
to make them capable of burning coal; all new industrial
facilities would be required to install coal-fired boilers.
While conversion of gas and oil-fired boilers to coal is
desirable, it is not simply a matter of removing existing
apparatus and substituting coal-burning equipment. There
are many instances where it would be exremely difficult and
expensive, or even physically impossible, to replace existing
oil or gas boilers with coal-burning facilities. For
industrial-size units, the cost of coal-burning boilers
and necessary support facilities far exceeds that of oil
and gas burning units; and in conversion, an added cost
allowance must be made for expensive pollution control
equipment.

On the supply side, a major constraint on the rapid
expansion of coal consumption is limited availability of
bulk transportation facilities. Present alternatives include
railroads, slurry pipelines, barges, and trucks, with most
shipments made by trains. Expansion of railroad capacity
should not be limited by the number of new hopper cars
required, since excess capacity exists in the plants that
manufacture them. Moreover, railroads would likely have
little problem raising capital to build hoppers. The
limitation in transport capacity derives from the poor
condition of railroad track beds. Except for a few major
routes, expensive construction and repair of track beds
and bridges would be required; however, no long-term delays
would likely result from this shortcoming.

Coal slurry pipelines could provide bulk transport
capability competitive with railroad transport. Under this
system, coal is crushed, washed, mixed with water and
pumped through a pipeline to be separated from the water
at the pipeline terminus. Unless expensive closed loop
systems are developed, such pipelines will be impractical
in some coal-rich areas where water supply is limited.
Railroads, of course, dispute the wisdom of building coal

slurry pipelines and object to having pipelines cross their
rights of way.

Expanding barge production appears to pose little
problem, but some industry officials note deficiencies in
lock and dam facilities. Increased truck transport is
not a problem except for its impact on rural road systems.

A major continuing uncertainty in the coal industry is
the future of labor relations. The recent coal strike, which
lasted a record 110 days, is a reminder of the element of
risk attached to dependence on coal as a fuel source. While
such strikes may have some chilling effect on expanded use
of coal, there is little indication that they will slow the
nation's demand for this commodity.

Insiders from the United Mine Workers Union are quick
to point out that it is not a one-man union. Not even in
the days of the formidable John L. Lewis was the union membership of one mind on many issues. Coal miners, who are
highly individualistic, are usually found in the forefront
of the labor movement, so it is unlikely that they could be
manipulated by any one person or group. The miners'
rejection of terms agreed to by union officials during the
1977-78 strike is evidence of this independent spirit.

Despite the length of the recent strike, many major
issues that led to that bitter contest have not been settled.
Union demands included wage increases and complete restora-
tion of previously reduced health, welfare, and pension
benefits. On the other side, management was looking for
establishment of penalties for wildcat strikers. Neither
side won. One panelist made several interesting observations
about that dispute which are significant for the future of
the U.S. coal industry. For example, during the last week
in February 1978, 8,000 miners were out in eastern Kentucky,
yet 1.4 million tons were mined that month, compared to 1.5
million tons produced the corresponding month in the previous
year. These statistics are indicative of a slide in the
UMW's power in recent years. Union coal used to account for
70 percent of the total, but presently this figure is down
to about 50 percent. Competition comes from the small "mom
and pop" mines that are not unionized. This competition is
one variable that the UMW must recognize in planning future
demands from mine owners.

As far as future expansion of the industry is concerned,
the labor force is not expected to be a constraint. It is
estimated that by 1985, 214,000 new miners will be required
to meet production goals, but company and union officials
see little problem in increasing the labor force to that size.

Several labor-related issues are now of concern to both
miners and mine operations. Chief among them is the level
of productivity. Wildcat strikes have cut into productivity
in recent years. The lessened average level of experience

in the work force has also contributed to the productivity
decline.

According to one expert:

> Throughout the 1950s and 1960s, the industry had
> an aging, stable work force. The influx of new
> men was so sparse that they could be carefully
> integrated into a work force of 40 or 50-year
> olds.... If a man runs a continuous mining
> machine for 25 years, he runs it a lot better
> in his twenty-fifth year than he did in his first.
> So productivity would naturally go up. But, by
> the late 1960s all those men started reaching
> retirement age. Coal was booming; and the nature
> of the work force flipflopped completely. You
> could have gone into a deep mine in western
> Pennsylvania, and the junior man operating a
> motor in a mine would have 30 years seniority
> back in 1968. Go to that mine today and you will
> find a kid who's been in the mine for three and
> a half years, and he is probably the senior
> motorman. That change in the work force, in and
> of itself, is going to cause a lack of expertise
> in mining and declines in productivity.

Government regulation is another cause of decreased
productivity. Occupational safety and health legislation,
passed for protection of miners, has added recently to the
burden of production; and even more extensive regulation may
be forthcoming. It is often the case that legislation is
passed without its sponsors knowing its full impact. The
1977 Amendments to the Mine Safety Act, for example, are
expected by some observers to reduce productivity, but not
to increase safety for miners.

One area of regulatory action needing attention is
utility rate-making. One panelist suggested that rate-
makers should turn to marginal cost pricing. In simple
terms, this means having customers pay an increased rate
for increased electric generating plant capacity required
to meet higher demands. One specific example of marginal
cost rate-making is the use of "peak-load" pricing surcharges
during various times of the day. A large portion of the
excess capacity in power-generating plants is held in
reserve for known daily peak periods. By employing peak-
load pricing surcharges for such periods instead of flat
rates throughout the day, peak usage can be discouraged
and more efficient use made of any given level of capacity.

In some cases, government regulations have hampered
energy efficiency. Cogeneration is such an example. Cogeneration technology is available now, and it could

provide a valuable means of energy conservation; however,
in some places it is specifically prohibited. Conservation
through cogeneration and other means has the advantage of
less damage to the environment than the alternative of
increasing supply.
One expert recapitulated the need for better regulation
in this area:

> We have to move toward an increased use of
> marginal cost-pricing principles for the pricing
> of electricity, so that the user of electricity
> has a better signal than he has now as to the
> economic consequences of his decision to use more.
> If we price electricity based upon historical costs
> of capital equipment used to generate electricity
> rather than on current marginal production costs,
> he is not getting the proper signal, and all kinds
> of end-use technology will be put off; it will not
> appear economical when it really is.

> Regulatory delay is one issue that applies to expansion
of coal production as well as to many other facets of energy
production in the United States. One example was cited in
the discussion, as follows:

> Several weeks ago, we did some work that indicated
> that if you started from scratch, applied for a
> federal lease in the West, successively cleared
> every hurdle before taking the next step, and never
> put yourself in a position of making an illegal move--
> it could take as much as seventeen years to get a
> new Western mine into production. There are 25
> major permits required and many more smaller ones.

A streamlining of the regulatory process, with the elimination
of costly, low-value requirements, is clearly needed to help
revitalize the coal industry.
Over the years, government intervention has distorted
price mechanisms to the point that serious inefficiencies
have resulted in a number of sectors. Transportation is one
industry which has been perversely distorted. Regulation of
rail, truck, and barge rates has caused disassociation
between cost efficiency and price in this industry which is
so important to energy. If a reduction in imported energy
use is to be achieved through efficient coal use, transportation prices will have to reflect resource evaluation
and costs measured by opportunities lost. Pricing in
competitive markets is the best-known process for achieving
such efficiency.
Government can contribute in ways other than removal
of obstacles to coal use expansion. One needed measure is

promotion of new technologies to transform coal to other fuel
forms and to eliminate potential pollutants in the process. The
federal government is already supporting technology development
in this area; additional encouragement could be given, however,
using such devices as tax credits. Other tax incentives, for
example, for installation of conservation devices and to facilitate the shift to coal from petroleum and natural gas could
also be useful. As noted earlier, using coal must be made
relative desirable, for the shift to coal is not so much a
matter of supply constraint as it is demand constraint.

In summary, coal constitutes about 90 percent of U.S.
energy reserves; it can clearly play a major role in the energy
future. Before an efficient expansion of the industry can be
accomplished, however, numerous hurdles must be cleared, including environmental effects of burning greatly increased
quantities of coal, ecological damage caused by mine drainage
and surface mining, development of improved transportation
systems, and various labor issues. The federal government
needs to take the lead in reducing the uncertainty associated
with these issues. Regulatory delay has to be reduced and
regulations of all types need to be restudied and perhaps
many of them changed if they are serious obstacles to efficient
resource allocation in the coal market, as well as other related markets such as transportation, electricity, and petroleum.
By reducing uncertainty, government can increase the rate at
which investments are made in coal mining and conversion processing. Reducing uncertainty would also expand investments
in related systems such as pollution control devices and
transportation facilities.

Both industry and labor assert that expansion in coal
production is not a barrier to large-scale substitution of coal
for oil and gas in line with the Administration's plan. Success
of the plan depends upon removal of barriers to coal utilization, for demand constraint is the problem. Some problems are
of a regulatory nature and must be addressed by the Congress
and Executive. Further measures to spur new technology are
required, too, if coal is to supplant oil and gas in some
of their uses. The Administration has made coal the center-
piece of the National Energy Plan; given the right encouragement, it could fulfill that role.

*   *   *   *   *   *   *

SUMMARY STATEMENT BY THE
HONORABLE ALBERT GORE

This interpretive report of the Quadrangular Conference
session on coal's future, which I had the honor of chairing,
appears to me to raise significant issues and to permit public
attention.

PANEL VI

NEW FRONTIERS IN ENERGY TECHNOLOGY

"New Frontiers in Energy Technology" was addressed
by a panel chaired by Senator Harrison H. Schmitt
(R, New Mexico), who is Ranking Minority Member of
the Senate Subcommittee on Science, Technology,
and Space. The panel included Daniel Ahearne,
President of Total Energy Application Management
(TEAM); Edward E. David, Jr., President of Exxon
Research and Engineering Co.; Pierre Desprairies,
Chairman of the Institut Francaise de Petrole; and
Denis Hayes of the Worldwatch Institute.

NEW FRONTIERS IN ENERGY TECHNOLOGY

At this time, investments in energy conservation appear
to have a greater payoff relative to additional investments
in production. Low costs of hydrocarbon fuels until 1973
produced a market where conservation was not rewarded as it
is today, so present payoffs to conservation investment are a
result of a transition from that era of cheap energy. Despite
governmental policy which subsidizes energy use, the U.S.
public (along with the populations of other industrial nations)
has proceeded to conserve energy in response to market forces.

Technologies for energy conservation and production are
bound to change lifestyles in the industrialized world, some
technologies more than others. One panel expert's view on
this point is as follows:

> Solar technologies are going to be producing fewer
> lifestyle changes than most of the alternatives.
> Our life today is still one based primarily on
> petroleum and natural gas, which are inherently
> very decentralized energy sources. The bulk of
> the options that most people are talking about,
> when they are talking about dream technologies for
> the next one hundred years, are highly centralized
> options. The terms one encounters include satel-
> lite power, magnetohydrodynamics, thermonuclear
> fusion, synthetic-fuels facilities. I have a sus-
> picion that solar resources, if they are tapped in
> an appropriate decentralized fashion, will be sim-
> ilar in their impact on society to those influ-
> ences that have been a part of the petroleum era
> and, in fact, the automobile age. It will be a
> largely spread out society with a series of small,
> clustered communities as opposed to large, cen-
> tralized cities. People will be highly mobile and
> you will have a relatively high degree of control
> over your lifestyle in your own domicile.

Solar energy appears to have the long-run edge in replacing fossil fuels that form the bulk of today's energy sources.
Several different sources of solar power, direct and indirect,
are available to tap with appropriate technology. Some of
this technology is available today, but it is not net employed widely because of relatively low cost of fossil fuels.
Increased world energy prices, combined with elimination of
current national subsidies for the use of fossil fuels, will
eventually make existing solar technologies economically competitive. Moreover, new solar technologies are being developed that will increase the solar share of energy consumed.

Sources of solar energy include direct sunlight, hydroelectric power, wind power, and biomass conversion using
trees, plants, and biological waste. Estimates of the portion
of energy consumption obtainable in the near future from solar
power range as high as 25 percent, a figure that includes
hydroelectric energy now being produced and exploitation of
existing dams presently without hydroelectric capability.

One process for tapping the sun's energy is direct conversion of solar radiation into electricity through employment
of photovoltaic cells. Large-scale use of this source depends
upon development of a storage system for energy thus collected,
and more importantly, upon improved conversion devices. The
solar cells now in limited use are made from silicon, and the
level of its conversion efficiency is relatively small. Since
present costs of producing cells are high, economical application of this technology awaits new and cheaper production
systems or an increase in cell efficiency.

The alternative problem of finding a means of energy
storage results from the intermittent nature of the energy
source. Daytime collection of energy only on sunny days
leaves a gap in required electric service. Of course, one
method of solving this problem is to use solar power in con-
junction with alternative generating equipment such as nuclear
or coal-free plants. However, the economical employment of
such plants depends upon a high-load factor, thereby preclud-
ing their use in a standby role. One possible means of stor-
ing solar energy is the pumped storage approach already
employed by some hydroelectric power stations. Under this
system, electrical energy produced by photovoltaic cells is
employed to pump water up to storage pools for later use in
hydroturbine generation. Alternatively, photovoltaic electricity might be employed to generate hydrogen from the electrol-
ysis of water. The hydrogen could be burned later.

Another system for using solar radiation that would not
be subject to constant disruption would employ photovoltaic
devices aboard large space stations placed in stationary orbit.
Electrical energy produced in this manner would be transmitted
to earth by microwaves where it would once again be converted
into electricity. Such a system would have the advantage of
placing photovoltaic cells in an unobscured path of the sun's
rays, beyond the earth's atmosphere. At this juncture, its
chief disadvantages appear to be a high capital costs and
unknown health hazards associated with exposure to intense
microwave radiation.

At one time, use of solar radiation was relatively popular in sunbelt states in the United States for limited purposes such as hot water heating. Devices like those now
being installed on house rooftops for this purpose, however,
were eventually driven off the market in the 1950s by cheap
natural gas. The rising costs of natural gas and oil over
the last few years are bringing thermal collectors back into
competition. Here again, problems of storage for use during

extended periods of cloudy weather still exist, so backup
systems are necessary.

The overall advantages of solar energy are its nonpolluting and nondepletable characteristics. Consequently, at a
time when stability in energy supply and quality of the environment are key issues, solar energy looks attractive as an
alternative as well as a follow-on to fossil fuels. The federal government is constantly urged to increase funds committed to solar energy on the grounds that a sufficient level
of advance has been achieved to warrant development of full-
scale solar power plants of various types. Experts seem to
agree that what is needed are not dazzling new technologies
but development of current technologies to economically competitive levels.

Nuclear energy promises to provide relatively low-cost,
pollution-free energy in the next few decades, but numerous
stumbling blocks exist to achieving its full potential. Different types of nuclear reactors are in various stages of
development throughout the world, and different types may
dominate the market in various places and at various times.
Not all nuclear development policy throughout the world's
governments is similar or necessarily compatible. The various
types of problems--weapons proliferation, radiation hazards,
waste disposal--are considered in a different order of priority by nations capable of nuclear research and development.

More than 60 light-water nuclear reactors are now in
operation in the United States, supplying about 10 percent of
electricity requirements. However, despite their excellent
record of safety and reliability, fuel supply problems have
emerged to dim an otherwise bright future. Recent estimates
suggest that the amount of uranium ore available for exploitation in the United States may be sufficient only to fuel about
600 standard-size light-water reactors for 40 years. Such a
level of reactor operation could be achieved early in the
next century, thereby creating a question for U.S. policymakers as to whether other forms of nuclear power--specifi-
cally the breeder--should be pursued. The Carter Administration has been ambivalent about the development of breeder
reactors, which create more fuel than they consume. The
Administration's concern is that plutonium extracted from
used breeder fuel rods during reprocessing could be diverted
for use in nuclear weapons. It is attempting to eliminate
the Clinch River Breeder Reactor program, but the Congress
does not concur; as a consequence, the fate of this option is
unclear.

Such is not the case in other industrial nations. In
energy-starved Europe and Japan, government policy has been to
pursue breeder research as a partial relief from a heavy dependence on petroleum imports. In France and the United Kingdom, in particular, there is wariness of the proposed U.S.
policy to retard breeder development. Were they to interrupt

their own breeder work for very long they could find themselves dependent on foreign goodwill and management for nuclear fuel supplies, much as they are now for oil.

A breeder reactor produces more fuel than it consumes and expands the usable energy value in natural uranium by 50 to 60 times, thereby providing a source of energy independence for uranium-poor countries. Research on breeder reactors was undertaken in Europe by a group of electric utility firms several years ago, and in 1974, Phenix, a 250-megawatt French breeder began producing electricity. Between 1974 and 1976, it maintained a 60-percent load factor, which is considered excellent for a prototype. Now France is building a 1,200-megawatt breeder called Superphenix. While construction costs of this large plant and other breeders will exceed that of light-water reactors, their economic competitiveness stems from efficient fuel utilization and the generation of fresh fuel. Several breeders combined with an array of light-water reactors are planned for the French power industry.

Several technological problems have had to be solved with development of the Phenix reactors that will be beneficial to breeder development worldwide. Among the problems addressed by this pioneering program is cooling the relatively hot core. Since the core is small, large amounts of heat generated by the fuel rods must be removed rapidly; but the coolant cannot be one that is a neutron moderator. Molten sodium is excellent for this process, and a system has been designed with a second-phase sodium heat exchanger to provide for additional protection against accidental reaction between radioactive sodium and the water used for the steam generator.

Regardless of the present policy of the United States regarding development and spread of breeder technology, the perceived need to pursue this technology as a means of economizing on scarce uranium resources will assure continued work on it outside the United States. France, Britain, West Germany, Japan, and several other Western nations are pursuing breeder development and it is clear that the United States cannot hold back this trend.

One alternative approach has been proceeding quite successfully in Canada. For more than 20 years, the Canadian government has been developing a reactor--called CANDU--using natural uranium as a fuel and deuterium as a moderating agent. The heavy water reactor has several advantages: first, the natural uranium used as a fuel is considerably cheaper and more plentiful than the enriched uranium used in U.S. light-water reactors; second, the use of individual pressure tubes holding the fuel provides an additional measure of safety against failure in the system which could do less damage than a similar failure in large pressurized vessels of the type employed in light-water U.S. reactors. In addition, such a design permits refueling while the reactor is in operation, thereby achieving greater efficiency through more

frequent removal of accumulated fission products without
reactor shutdown or significant loss in power on output. As
a result of these advantages, today this system is operating,
or under construction, in India, Pakistan, South Korea, and
Argentina.

One further form of nuclear reaction that promises an
almost endless supply of energy in the distant future is that
of fusion. Fusion is the source of energy released on the
sun; it has also been employed on earth in the form of the
hydrogen bomb. Fusion's advantage is that it can be achieved
with deuterium, the isotope of hydrogen found in heavy water,
which is relatively abundant in sea water and can also be
manufactured. Its disadvantage is that the circumstances
necessary for the controlled release of energy are so limit-
ing that the technology to achieve it has not yet been perfected, and may not be for many years. To create fusion
reaction, temperatures of about 100 million degrees Celsius
must be reached, and technology to control such heat is still
perhaps 20 years in the future. To put the present state of
fusion development in perspective, one panelist noted that
by 1983, fusion technology will be at the same level that
fission technology was in the nuclear reactor at Chicago in
1942. No country working on fusion--the United States, USSR,
and a European group--has yet progressed very far.

Starting and then maintaining a controlled fusion
reaction inside a chamber are two key technological problems.
Ignition research has concentrated recently on using lasers
to achieve the needed high temperatures. Several high-
powered laser beams are simultaneously focused on a target
area containing fuel, generating a small explosion which
produces a plasma. Plasma is a state of matter composed of
the electron-free nuclei of various atoms in an extremely
active suspension. Fusion takes place when the nuclei fuse
together, due to high speeds, producing heat that can drive a
steam generator.

Controlling plasma in a contained environment is the
second major hurdle. Temperatures of the needed 100 million
degrees Celsius would melt any material available on earth.
One method for lining a fusion reactor is through use of a
magnetic field. The most effective designed so far using
this approach is the tokamak system developed in the USSR
and now employed in fusion research in several countries, including the United States. Success has thus far been limited,
and a great deal more research in plasma physics, laser bom-
bardment, and other areas will be required before fusion
power generates any electricity.

Coal is one of the most abundant sources of energy in
the world, and given present levels of technology, a variety
of research efforts are now under way to find additional and
more flexible uses for it. Synthetic fuels from coal are
becoming a reality in several nations, and the United States
promises to be in a preferred position (with coal making up

the vast bulk of its fossil fuel reserves) to take advantage
of new technologies.

Of course, production of gas and liquid hydrocarbons
from coal is not new. In the decades before cheap natural gas
was available from the oil fields of the Southwest, gas production from coal was common in many cities and towns. And,
during World War II, German technology produced a method of
making oil from coal. Several processes are now available
for synthetic oil and gas production, including carbonization,
hydrogenation, liquefaction, and extraction procedures. Any
one or a combination of these processes may turn out to produce synthetic fuel at competitive costs.

In any probability, production facilities for synthetic
fuels will be located near large coal mines in order to reduce
transportation costs, since anywhere from 25,000 to 100,000
tons of coal would be consumed in one day's operation of a
plant large enough to produce at a competitive cost. Since
large plants have never been successfully demonstrated, and
because the capital cost of such a facility could run into
billions of dollars, a great deal of risk is involved for
private firms contemplating a start-up. In order to overcome
the initial costs of demonstration, synfuel supporters in the
United States are asking the federal government to provide
capital for plant construction. Private industry could op-
erate facilities and eventually purchase them from the government, but industry is unlikely to proceed on its own because
of risks and costs.

In addition to the fuels already cited, there is biomass
conversion in which garbage or forest or farm products are
converted into synthetic fuels. Another alternative is mag-
netohydrodynamics where, for example, coal or another hydrocarbon is employed to produce electricity directly from a
magnetic field. These options and others are now being
tested and perfected in public and private laboratories
throughout the United States and may someday provide a significant share of our fuel supplies.

Progress in new energy technologies is hampered by in-
sufficient basic research. In the United States overall
national expenditures on basic research have declined in real
terms in the last few years, especially by the federal government. Private enterprise has chosen to pursue development of
old technologies more certain to produce payoffs in the near
future. University research has declined; some observers feel
this is because the universities have too little long-term
support and spend too much time raising funds. In addition,
computation equipment and instrumentation in many schools are
antiquated because of lack of capital.

Meanwhile, we seem to have been moving into an anti-
technology mood, at least in many quarters, in recent years.
Much of this bias appears to be reinforced by a distrust for
large organizations such as big business and government, which
are perceived as the "keepers of technology."

The potential of basic research can be appreciated by
looking at one example, "surface science." This is not a new
science, but one that has taken advantage of new procedures
and equipment such as scanning electron microscopes. Understanding surfaces can help scientists produce lubricants which
reduce friction and aid in energy conservation, or surfaces
may be keys to improving solar collection by increasing energy
absorption and reducing reflection. Only by increasing our
commitment to such basic science as well as maintaining technological developments necessary to exploit success in basic
research can we create new alternatives to continued reliance
on petroleum.

In contrast to declining support for basic research as
a whole, in the last few years the amount of money made available by the government for energy research, development, and
demonstration has increased considerably. In 1975 Congress
created the Energy Research and Development Administration
(ERDA) to supervise and coordinate energy research funded
from Washington. Despite this increased emphasis, various
critics argue that R&D money has been too slow in coming.
Solar advocates point out, for instance, that the federal
solar energy budget is only the equivalent of funding to produce four B-1 bombers, despite solar technology's promise as
a nondepletable, nonpolluting energy source.

Many analysts have argued that the most efficient way
to encourage energy research and development would be to re-
move price controls from domestically produced oil and gas.
Higher prices would provide incentives for increasing attention to alternative energy and conservation technologies.
The suggestion has been made, too, that decreased regulation
and direction by the government in the technology area would
be helpful. These complaints that governmental direction of
R&D often discourages pursuit of technologies that might
prove useful appears to conflict with other voices urging
further concentration in support for specific lines of research. Obviously, both suggestions cannot be adopted, given
ever-present budget constraints.

Government support for various new technologies gains
some credence from the fact that some on-line energy systems
are presently receiving subsidies. It was pointed out that
the nuclear power industry, for example, receives subsidies
in the form of government-backed liability insurance and
federally-funded reprocessing plants. So the argument was
that some new, presently "marginal" energy systems will only
prove economically viable if they, too, receive federal support of one kind or another. A related argument for governmental intervention is that public funds should underwrite
new institutions to sponsor energy development and investment
in energy-conserving equipment.

Diversity should be a goal in governmentally-supported
R&D programs, according to some analysts. They emphasis
diversity of needs within the complex societies of the West

and suggest that R&D policy should encourage pursuing alternatives just as variegated. Unfortunately for supporters of
this view, the centralized nature of government decision-
making tends to limit certain alternatives. In those cases
where advocates of a single approach are influential, for
example, priority may be assigned to such an alternative to
the detriment of others as promising or more promising. In
any event, policymakers in Washington can never recognize
fully the special requirements in different places at different times, and diversity tends to be lost in the shuffle,
regardless of how committed to diversity the policymaker may
be.
     One expert arguing on behalf of diversity put the case
this way:

     What is required is a portfolio of diverse possibil-
     ities, so that we can accommodate what is bound to
     be the unpredictable. In many ways this principle
     of diversity may seem at odds with the urgencies
     that are being pressed on us about the energy situa-
     tion. And yet I believe that we should not con-
     strain our technological choices too soon and
     should not mastermind the situation too early be-
     cause there are possibilities and needs which we do
     not yet see.

For the next few decades the scenario will probably be
one of moderate change in energy technology. Another panelist
evaluated the near future this way:

     We will have to live with fossil fuels comple-
     mented by improved technologies, not revolutionary
     technologies. In the field of oil and gas we will
     have to develop enhanced oil recovery technology
     to extract around 40 percent and not 25 percent
     as we do now. We will have to develop deeper sea
     exploration to a greater extent. For coal, I am
     very confident about its development through the
     new technology of gasification; however, I am
     afraid that coal will be, in the meantime, only a
     substitute for nuclear energy, and not much more
     than that.

     In summary, the future of energy technology now depends
heavily on the money and the wisdom of governments. Since the
cost of energy research is borne in considerable part by tax-
payers, special interest groups are increasingly pressing their
cases for additional financial backing from public coffers.
In their turn, public officials are faced with the difficult
mission of trying to sort out promising research paths from
inappropriate lines of pursuit, a task that a centralized,
multilayered decision-making process tends to accomplish awk-
wardly and belatedly.

Some argue that creativity in energy research does not
have any connection with the government. One expert noted:

The real focal point is not today in Washington,
D.C., but rather it is down in the basements of
tinkerers, in small garages, and machine shops.
It is found in relatively small industries, capi-
talized at a few million dollars, where innovation
is coming off and where people are truly elated
by this potential that is out there to be explored.
Someone wrote months ago that Washington is the
hub of the country because everything else seems
to be moving.

Whether the lead in research and development should
come from the public or the private sectors, it is proceeding
in varying ways in both. More and better efforts are un-
doubtedly required in both if we are to adapt smoothly to the
certainly different energy future. We will have to await
the verdict of history to see how effectively present energy
research policy is performing.

*   *   *   *   *   *   *

SUMMARY STATEMENT BY THE
HONORABLE HARRISON SCHMITT

It is always gratifying to participate in debates on
great issues, and energy is truly such an issue. While I do
not necessarily subscribe to all viewpoints expressed by the
panel members, it is evident to all workers in the real world
of energy that we must produce to survive our present dependency on fossil fuels. This production will buy time until
we have fully converted from a "dependent" energy society to
an "independent" energy society. Conferences such as this
serve the very real need clearly and expertly to focus attention on what our future can be. Meanwhile, those of us in
government must fight our way out of the mistakes of policies
that have artificially restricted the growth of new, more efficient energy technologies.
An energy policy for Americans must be positive in its
spirit. It must call on the best in the American character--
the creativity, the competitiveness, and the compassion that
conquered American wilderness and met the challenges of the
last 200 years. Most of all, our energy policy must rapidly
give young men and women the motivation, the resources, the
information, and the leadership they can use to move the
nation from a time of impending energy crisis to a time of
energy balance and growth. This is the way our challenges
have been met before; this is the way the energy challenge
can be met now.

PANEL VII

SECURITY DIMENSIONS OF THE ENERGY PROBLEM

"Security Dimensions of the Energy Problem" was
the subject of the final session of the Fourth
Quadrangular Conference. The session was chaired
by Dr. Amos Jordan, Executive Director for Inter-
national Resources Programs of the Georgetown
Center for Strategic and International Studies.
Members of the panel included Norman Chappel,
Minister Counselor for Energy of the Embassy of
Canada; Melvin Conant, President, Conant and
Associates, Ltd.; Robert Ebel, Vice President of
International Development, ENSERCH Corporation;
Admiral Thomas Moorer, former Chairman of the
Joint Chiefs of Staff and now member of the Cen-
ter's Advisory Board; and Lt. General William Y.
Smith, USAF, Assistant to the Chairman, Joint
Chiefs of Staff.

SECURITY DIMENSIONS OF THE ENERGY PROBLEM

In considering the security dimensions of the energy
problem, a picture emerges of a complex relationship that had,
in the past, been inadequately considered by government. International developments in the last five years, however, are
forcing governments to address how energy problems influence
national security and vice versa.

Four interrelated concerns emerged from the discussions.
First, governments recognize an absolute need for an adequate
world supply of energy, especially oil. A shortfall in world
oil supplies--whether politically contrived or resulting from
technical or physical reasons--would be disastrous for most
states. A recent Exxon study estimated that by 1990 free
world oil demand will reach more than 72 million barrels per
day (mbd). To meet this demand, Exxon estimated that members
of the Organization of Petroleum Exporting Countries (OPEC)
would have to produce about 44 mbd. Recent studies by EPRI
(The Electric Power Research Institute) are a bit more reas-
suring; they estimate total demand at 64 MBDOE by 1990 with
OPEC's needed supply at 36 MBDOE. Whether OPEC will produce
at either of these levels is a key question. Will Saudi
Arabia, in particular, expand its production capacity (now
about 9 MBDOE) sufficiently to provide roughly half (18-22
MBDOE) of the required supply? There is little assurance
that it will.

Second, even if there is a sufficient overall supply of
oil, the continuity of flow of oil must be assured. No
country could withstand for more than a few weeks (or months,
even if it has an ample stockpile) the severe negative impact
of a major disruption of the flow of oil from producing countries in the Middle East and elsewhere in OPEC.

Third, the price of oil should be such that countries
can pay for it as part of normal trade, not by accumulating
huge debts or through special deals such as swapping oil for
arms.

Fourth and finally, every country is concerned about the
political and military stability of the Middle East that is,
and will remain, the preeminent source of the oil in world
trade. Other regions will not be producing sufficiently by
1985 or 1990 to dilute the overwhelming importance of Middle
Eastern oil. Saudi Arabia, in particular, will likely assume
an even more important position in the oil market than it now
has. The location of so much oil in such a volatile area as
the Middle East can only continue to generate anxiety among
oil-importing states.

These four concerns are shared by all oil importers.
Developing countries, however, have additional problems.
Most less developed countries (LDCs) have political and eco-
nomic modernization as paramount goals, and if they are to

industrialize, they must base industrialization largely on
energy from petroleum. In the near term, at least, they cannot escape the petroleum age. If oil supplies tighten and
prices increase significantly in the mid- or even early-1980s,
as analysts expect they will, some LDCs will be caught in
the middle of their industrialization, others will still be
at the beginning; none will be immune. How will they be able
to compete for limited supplies available on the world market?
It is difficult to imagine that LDCs facing tight supplies
will be able to compete effectively unless both oil producers
and major consumers of the industrialized West make conces-
sions regarding both price and supply.

For the industrial world, the situation is somewhat different. There the concern, particularly in Europe, centers
on anticipation of shortages leading to competition, not
between oil companies, but between governments. Such competition could lead to serious strains with NATO, with negative
repercussions for all members.

Among Western countries, the United States stands apart
in that it has not had long experience with dependence on
foreign sources of energy as have Europe and Japan. Two
consequences flow from this: first, the Europeans and
Japanese often perceive an over-concern in the United States
about dependability of supply; these allies consequently tend
to see American reactions (e.g., proposals for breaking OPEC)
as more inflammatory than need be, making resolution of differences more difficult; second, from the European and
Japanese perspectives, the major question during the summer
of 1978 was whether the United States would develop an energy
strategy commensurate with its predominant position in the
oil market. Today the United States imports 20-25 percent
of all oil in world trade. For too long, American executive
and legislative policymakers have not shown that they are
serious about reducing this overblown role.

In addition to its status as a massive consumer, there
are additional implications for the United States as a super-
power and leader of the West, when questions are raised regarding overall supply and dependability of oil. Obviously,
the United States must work closely with its allies to ensure
adequate energy supplies for all. The International Energy
Agency (IEA) represents a major step in this endeavor. In
the IEA, the industrialized democracies have already reached
agreements for an emergency oil reserve and allocation system.
They have also agreed to reduce oil imports and implement
measures to spur development of alternative energy sources.
All of these measures are intended to reduce the vulnerability
of the West to disruptions of oil supplies.

In general, the IEA--stemming as it does from a common
concern with security of energy supplies--could become an effective instrument for allocating responsibility for energy
matters and a forum for broader international cooperation in
the energy field. It certainly represents a major step

forward from earlier efforts such as the energy committee of
OECD. Nevertheless, to be completely effective, the IEA will
require continuing "credible efforts" by its members. Linger-
ing doubts persist, for example, about the political will of
IEA members to execute their present agreements. Such doubts
cannot be dispelled overnight; years of practice and commitment will be required. The United States, in particular,
will have to strengthen its leadership in all aspects of the
IEA if that organization is to become an instrument by which
the four concerns of Western oil importers outlined earlier
can be dealt with effectively. One American-IEA initiative
that was urged on the panel was the convening of periodic
consultations among the key oil producers and large importers.
It was argued that until this is done, none of the problems
associated with security of oil supplies can be seriously
addressed.
     One question of importance to all consumers is the fu-
ture position of the Soviet Union in world oil markets: Will
the Soviet Union become an early net importer of oil as CIA
estimates have predicted? If the Soviet Union becomes a net
importer rather than--as now--an exporter of oil, what will
be its impact in the Middle East and elsewhere? According to
one expert,

          The status of Soviet oil is now not all that dif-
     ferent from the situation of the United States in
     the late 1960s, when we advised our OECD colleagues
     that our spare producing capacity had been ex-
     hausted, that we could no longer be looked to for
     additional supplies of oil in an emergency, and
     that our dependence on imported oil was very apt to
     increase, with attendant political and economic
     ramifications.

     Several problems confront the Soviet Union in oil and
gas production. Production in older fields has peaked, and
declines in their outputs are beginning to accelerate. Moreover, new finds have been made relatively slowly. Serious
difficulties have been confronted in both oil and gas production in Western Siberia. This region has emerged as the
primary alternative to the older Urals-Volga area, whose oil
fields had been the center of the Soviet oil industry for
the first two postwar decades. Most major Siberian gas
fields, for example, are north of the Arctic Circle where
severe weather and permafrost make exploration and production
difficult. In addition, the region suffers from a lack of
manpower, particularly of skilled workers. For those people
who have gone to Western Siberia, poor living conditions and
the absence of an infrastructure have diluted the advantages
of higher rates of pay. Consequently, Western Siberia is experiencing high labor turnover.
     Despite these problems, Soviet planners have decided
that efforts in Western Siberia must be intensified.

Increased production there represents the only short-term
solution to U.S.S.R. oil supply problems. Coal and nuclear
power are not now feasible alternatives. The search for new
oil in deep-seated formations of established regions, in
offshore waters, or in unexplored territories, can only have
an impact in the distant future. Only through greater efforts
in Western Siberia, and perhaps through a reduction of oil
exports--first of exports sold for hard currency and then of
those sent to the East European satellites--can the Soviet
Union buy the time it needs to develop its long-term oil
potential.

It was argued that the Soviet Union will not accept
"the political and economic vulnerability" brought about by
imports of large volumes of energy. If it does not become an
importer, Moscow seems unlikely to share the major concerns of
non-communist oil importers outlined earlier. Given this
assumption, we should ask to what extent Moscow will feel
constrained in exercising a disruptive influence in world oil
trade or in oil-producing regions such as the Middle East.
Past records suggest that the Soviet Union will probably not
be disruptive, but the past may be a poor guide to the future,
particularly if Moscow sees opportunities in the Middle East
at a time when it is experiencing serious oil production
shortfalls.

Whether or not the Soviet Union becomes more involved
in the Middle East, the region--Saudi Arabia in particular--
will be of prominent concern for all oil importers. A continued massive flow of oil from the area is vital, and any
regional instability--whether internal or among countries of
the region--can jeopardize that flow. Unfortunately, there
are ample opportunities for instability. Large injections of
oil money into Middle East producing states have accelerated
their modernization processes and exacerbated their short- and
medium-term social and political problems. This is as true
of Iran as of its Arab neighbors; serious tensions are present
in both. Moreover, traditional rivalries among countries of
the area threaten further cross-border tensions, or hostilities.

The most visible dispute of major concern, of course, is
between Israel and her Arab neighbors. A further outbreak of
hostilities between the Arabs and Israelis might affect the
oil flow from the Middle East in several ways. First, oil
production or transport facilities could be damaged during
hostilities. The growing number of pipelines through the
region make attractive targets. Second, as happened in 1973-
1974, an Arab oil embargo, shutting off or curtailing the flow
at least until well after hostilities, might well be instituted.
How far an embargo would extend beyond the end of a war would
depend in part on the conflict's outcome. In 1974, for example,
the partial embargo was not lifted until the United States
demonstrated that progress was being made in negotiations between Israel and Egypt over troop withdrawals. Third, a
serious Arab reversal in a conflict might result in the over-
throw of moderate regimes that have taken a mild line on oil
prices and supply increases.

Although the Arab-Israeli dispute has high visibility, several other problems could also generate instability in the Middle East. Concern was expressed, for example, over differences between Saudi Arabia and Iran, and whether those differences can be managed. One major concern arises out of the growing seriousness of the revenue problem in Iran, and lack of such a problem in Saudi Arabia. The existence of huge oil supplies in Saudi Arabia and its lesser neighbors, just across the Gulf from Iran, could prove tempting and lead to conflict. In such a situation, the Western powers would confront an extremely difficult problem. Would the United States try to manage the crisis alone? How could the Europeans and Japanese become more involved? How could the U.S.S.R. be kept from exploiting such a contingency?

Iran also has a traditional rivalry with Iraq. Iran's support for Kurdish insurgents in Iraq, dispute over the Shatt al-Ahrab at the Iranian-Iraqui border, and ideological disagreements have caused serious friction, and occasional conflict, between the two countries. Recently, however, Iran and Iraq have patched up their differences. Termination of Iran's support for the Kurds and a settlement of their boundary dispute have set the basis for increased cooperation. But even if tensions within Iran and Iraq do not boil over and the Kurdish and Shatt al-Ahrab differences between them remain submerged, the two are natural rivals. Future developments could well undermine the new-found spirit of cooperation. Since relations between Iran and Iraq are a major factor in defining the stability of the whole Persian Gulf region, they bear careful watching by Western policymakers.

Industrial oil importers must be sensitive to developments throughout the Middle East, as well as to their own relations with states of the area. One expert argued, for example, that the interests, views, and needs of oil-producing countries must be taken more seriously into account in formulation of overall foreign policies--not just energy policies-- of the Western oil-importing countries.

In general, too, all states must become more aware of the security dimensions of energy issues. Even after the 1973-1974 oil embargo, there was reluctance--at least in the United States--to draw a relationship between energy problems and security concerns. It is only in the last two or three years that American policymakers have become sensitive to the links between these issues.

Growing awareness in the United States of a security-energy relationship has manifested itself in several ways. In official statements of American energy goals, for example, there is emphasis on *secure* access to energy for the United States and its allies. Prevention of a takeover of oil-exporting states by unfriendly forces is also now articulated as a primary defense goal. From the Pentagon, Secretary of Defense Brown has stated repeatedly that one of the most serious threats to national security is the heightened insecurity of

energy sources and the absence of assured sources of supply.
In Presidential Review Memorandum 10 (PRM-10), the Carter
Administration's major review of national strategy made public
in September 1977, it was stressed that the United States has
to be prepared for minor contingencies. *For the first time,*
such a contingency in the Middle East and Persian Gulf areas
was spelled out.

Secretary Brown subsequently followed up the PRM-10
statement. In order to prepare for a possible threat to oil-
producing facilities in the region, he declared that three
U.S. divisions will be earmarked as a quick reaction force to
respond to crises in regions outside Europe such as the
Persian Gulf. Planning for this force is reportedly already
under way. Backbone of the force will reportedly consist initially of the Army's 101st Air Assault and 82nd Airborne
divisions, which are maintained in a high state of readiness
for rapid deployment overseas. It is envisioned that in
periods of tension these forces will be able to deploy rapidly to the scene of a crisis--in order, for example, to preempt
action against the Gulf oil fields. Some estimates indicate,
however, that a really effective quick reaction force will
take several years yet to develop, owing largely to a current
lack of adequate airlift capacity.

Security of oil supplies at the source, i.e., preserva-
tion of facilities so that producers can keep the oil flowing,
is only the first dimension of the problem. Since the oil
must be transported to where it is used, supply routes must
also be secured. The problem is particularly acute during
wartime when oil would be necessary not only to fuel industrial production supporting a war effort at home, but also to
power the armed forces at the scene of conflict or crisis.
Recent trends have generated heightened concern among security
analysts about whether oil supply routes to the West are now
secure.

There are four primary sea lanes of communication (SLOCs)
carrying oil from the Persian Gulf--through the Suez Canal,
around the Cape of Good Hope, through the Indonesian Straits,
and around Australia. Of these four, only the route south of
Australia is relative invulnerable.

In recent years, the vast majority of oil from the Persian
Gulf has been shipped around the Cape of Good Hope. Between
1965 and 1977, the amount of oil transiting the Cape increased
from 800,000 barrels per day (bpd) to 18 million bpd. The
reasons for this increase were heightened demand for oil, the
closure of the Suez Canal in the wake of the 1967 Arab-Israeli
war, and much greater use of supertankers. Between 1971 and
1976, ships of over 205,000 DWT constituted more than 72 percent of the tonnage built. Ships of this size are too large
to pass through the Suez Canal and must go around the Cape.
They have helped to make the Cape Route the busiest sea lane
in the world.

Western security analysts, then, are most concerned about
security of the Cape Route. Japanese defense planners, of

course, have a quite different problem, one to which the
United States must also be sensitive, namely, assuring transit
of Persian Gulf (probably around Australia) and Southeast
Asian oil. Growth of oil trade around the Cape, and through
the Indian Ocean more generally, has coincided with a second
trend that has heightened anxiety about vulnerability of sea
lanes. That second trend is growth of Soviet naval power in
the Indian Ocean region. The Soviet Union was quick to insert
a maritime presence into the area in the late 1960s once the
British had announced withdrawal. They have maintained a
naval presence in the Indian Ocean ever since.

     One panelist argued that the evolution of the naval
balance in the Indian Ocean must be viewed against the changing pattern of access to facilities in the region for the
United States and the Soviet Union. In recent years, the
United States has lost access to ports in most countries on
the Cape Route. With the Portuguese withdrawal from southern
Africa, for example, U.S. naval forces can no longer stop in
Mozambique, Angola, Cape Verde Islands, or Guinea-Bissau.
For political reasons, use of South Africa's excellent facil-
ities--particularly those at Simonstown--are also denied
(although during wartimes these facilities would probably be
used). In the region, the United States has access only to
Mombassa in Kenya, Diego Garcia and--in a limited way--to
Bahrain.

     In contrast, the Soviet Union has pushed hard for access
to facilities in the Indian Ocean region. Moscow currently
has the use of naval and/or air bases or facilities in Iraq
and South Yemen. (They have been building up considerably in
the latter since their expulsion from Berbera, Somalia.) In
addition, although they do not have bases in Angola and
Mozambique, Soviet forces also make use of these countries'
ports. Moreover, Soviet naval forces have anchorages at several points in the Indian Ocean, including the Seychelles and
the Chagos Archipelago (of which Diego Garcia is a part).

     Some Western security analysts argue that the Soviet
naval buildup in the Indian Ocean combined with Moscow's
political and military activities in the Middle East and Horn
of Africa are signs of a Soviet strategy designed to encircle
Middle Eastern oil. Developments in other countries, such
as the recent pro-Soviet coup in Afghanistan, would enhance
such a strategy, if it exists. Whether Moscow is consciously
pursuing such a design is not really the question; the results are the same: namely, that the Soviet Union is increasingly in a position to exert pressure against oil life-
lines from the Persian Gulf to Europe, the United States, and
Japan.

     It is not only in the immediate Persian Gulf region or
in the Indian Ocean generally that there is cause for concern.
The Soviet Navy also maintains large submarine forces in both
the Atlantic and Pacific. In times of crisis, these forces
would threaten shipping plying the Cape of Good Hope and
South Atlantic routes, as well as vital maritime commerce

between North America and Europe. They would also threaten
Japanese and U.S. SLOCs through the Pacific.

One alternative sometimes suggested to offset this
maritime threat, at least in times of crisis or conflict, is
an airlift. The American airlift of supplies to Israel during
the October War of 1973 demonstrated how critical such an
operation can be. In some cases, American cargo planes were
delivering ammunition to Tel Aviv in the morning that Israelis
were firing in the afternoon. Effectiveness of an airlift,
however, is limited to the short term and to high-value cargo.
During the 1973 operation, for example, for every ton of
supplies that was delivered to Israel, four tons of fuel were
used. Clearly, if petroleum were one of the supplies being
airlifted, the exercise would be foolish. There is no alternative to the transportation of bulk supplies by sea.

Major Western oil importers--most of whom are also
members of NATO--are letting themselves get into a strategic
position that could have fatal results unless they establish
a means to protect oil flows. If war were to erupt in
Europe, it would not necessarily escalate immediately to a
nuclear exchange. Nor can we assume it would be a short war,
lasting no more than 30 to 60 days. NATO defense planners
must plan for a longer war, and if it were to occur, Western
Europe and the NATO forces would need Middle Eastern oil to
survive and prevail. A successful war effort, then, necessi-
tates successful protection of the sea lanes from the Persian
Gulf.

Consequently, NATO can no longer be solely concerned
with events north of the Tropic of Cancer. This artificial
geographical boundary of NATO interests has constrained NATO
thinking for too long. While NATO is a North Atlantic
alliance, the international situation denies its members the
luxury of thinking solely in North Atlantic terms. It is
highly unlikely, however, that NATO members will make the
decision to extend the alliance's southern boundary. Such a
decision would require the unanimous consent of alliance
members and such consent is not likely to be forthcoming at
this time. There are reports, however, that NATO has developed plans for defense of the sea routes around the Cape.

Finally, defense policymakers must address the political
implications of hostile pressure against the flow of oil to
the West (including Japan in that generic term for the industrialized democracies) in situations short of war.
Whether that pressure is exerted at the source against oil
producers or against oil in transit, the possibility of an
effective threat to the oil flow creates perceptions of
Western vulnerability. If the West is perceived as unwilling
to alleviate this situation, states in the Middle East and
elsewhere will judge the balance of power to be shifting away
from the United States and its allies. The Middle Eastern
states are highly sensitive to security developments in their
region. At the very least, then, the question of the

security of Middle Eastern oil supplies has a major impact on East-West competition for psychological advantage that can possibly be parlayed into political influence.

Of course, the secured access-transport problem is not limited to the Middle East. What are the implications, for example, of Soviet naval bases in Cuba for oil flowing through the Caribbean? How secure are oil supplies from Nigeria and Indonesia? The U.S. Department of Defense is only now beginning to ask these and similar questions.

The relationship between energy problems and national security is complex and multidimensional. For now, the key to that relationship, however, can be summarized in a single phrase: "access to oil." If industrial states are to remain economically healthy and if they are to maintain their national defense, they must have continued and secure access to adequate energy supplies which, in the short term, means access to oil.

The need for secure access to oil supplies will have a tremendous impact on national foreign and defense policies. The requirements it imposes on those policies may well conflict directly with other foreign policy goals, as the dilemmas confronting U.S. Middle East policy sometimes demonstrate. The need for access should also contribute to selection of allies and to redefinition of threats to national security. The fact that many states which control oil resources are in the developing world adds a broader "North-South" dimension to the problem.

American foreign policymakers have only recently paid attention to the difficulties created in trying to reconcile energy policy and security policy objectives. Though the task confronting them will not be easy, it must be accomplish ed if the United States and its allies are to have adequate energy supplies and a strong national defense.

*   *   *   *   *   *   *

SUMMARY STATEMENT BY
AMOS A. JORDAN

The foregoing report reflects fairly the character of the discussion of the important linkage between security and energy, although it inescapably fails to capture many of the nuances expressed. I believe the principal linkages between security and energy have been raised, except for one that is less direct than those cited; I refer to the likelihood that an overly compressed transition, or an inadequately prepared transition, to reliance on other than fossil fuels will place such heavy economic strains on the industrial democracies that both their cooperation with each other and their internal cohesion will be in serious danger. The domestic and foreign policy strains induced by the short-lived embargo and

87

quadrupling of oil prices in the early 1970s were only a
foreshadowing of the stresses on international and internal
security that may be generated in the 1980s or 1990s.

Quite clearly, the West cannot afford to gamble that,
somehow, painlessly, its energy problems will be solved. Too
much is at stake and the required lead times for action are
too long to permit largely business-as-usual attitudes to
prevail. If we are to limit the danger of grave insecurity,
we should urgently develop *all* the alternatives to oil imports,
accelerate development and exploitation of new technologies,
emplace strategic stockpiles in all the industrial democracies,
strengthen bilateral and OECD relations with as many as
possible potential suppliers, and look to the adequacy of
our naval strength.

APPENDIX

MEXICO'S OIL
Trends and Prospects to 1985

by

Sevinc Carlson

May 1978
(Updated July 1979)

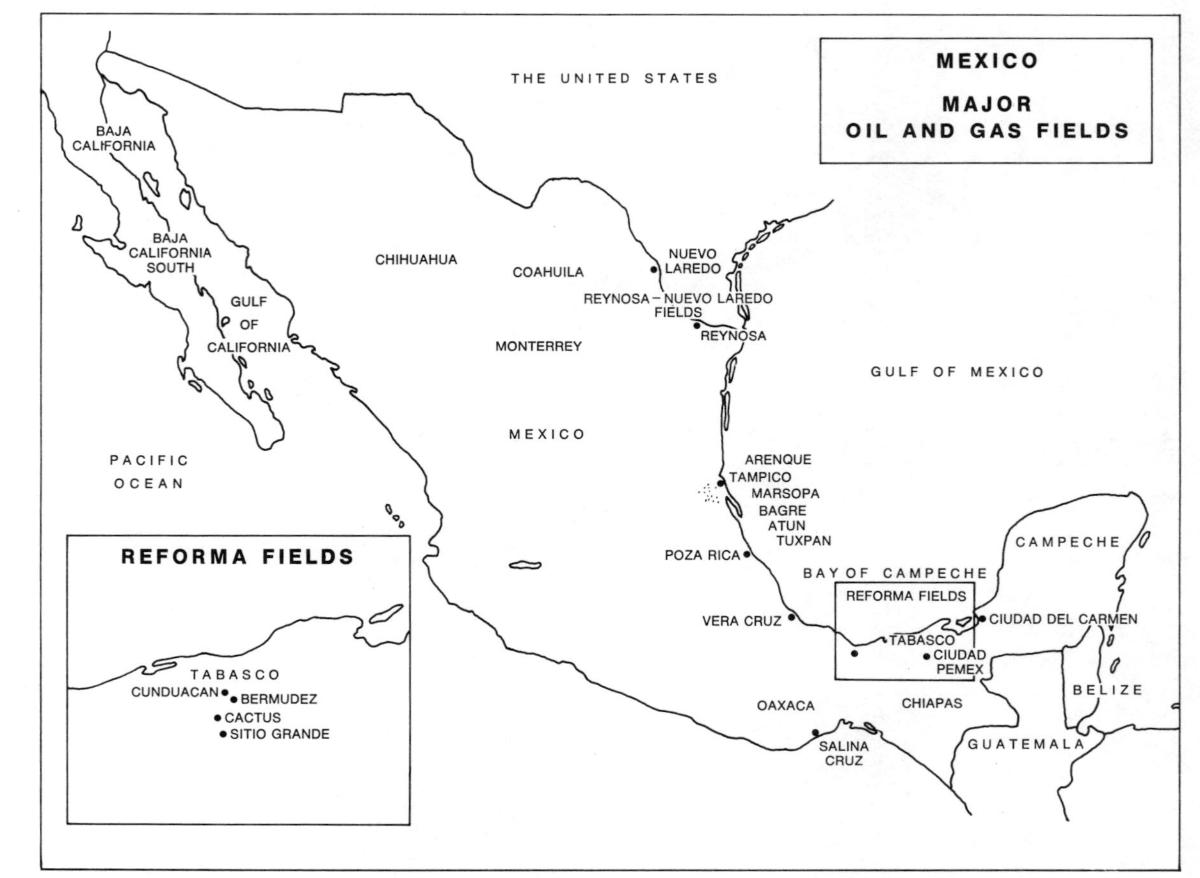

MEXICO
MAJOR
OIL AND GAS FIELDS
THE UNITED STATES
BAJA CALIFORNIA
BAJA CALIFORNIA SOUTH
CHIHUAHUA
COAHUILA
NUEVO LAREDO
REYNOSA-NUEVO LAREDO FIELDS
REYNOSA
GULF OF CALIFORNIA
MONTERREY
MEXICO
PACIFIC OCEAN
GULF OF MEXICO
ARENQUE
TAMPICO
MARSOPA
BAGRE
ATUN
TUXPAN
POZA RICA
BAY OF CAMPECHE
REFORMA FIELDS
VERA CRUZ
TABASCO
CIUDAD PEMEX
CIUDAD DEL CARMEN
CAMPECHE
BELIZE
OAXACA
SALINA CRUZ
CHIAPAS
GUATEMALA
REFORMA FIELDS
TABASCO
CUNDUACAN
BERMUDEZ
CACTUS
SITIO GRANDE

INTRODUCTION

Mexico's oil and gas reserves and its potential as an oil
exporter to the United States to help diversify the U.S.
oil imports, have been the subject of considerable speculation during recent years.

Mexico was one of the earliest oil producing countries
in the Western Hemisphere. Production of oil in Mexico
started before World War I. Mexican oil production, which
was largely controlled by U.S. oil companies, was considerably increased during World War I. During the early 1920s,
Mexico became one of the world's leading oil producers with
over half a million b/d* or about 25% of the world's supply.[1]
Most of this production was exported. Although oil production declined after the early 1920s, Mexico continued as one
of the world's significant oil exporters until 1938.

As a result of continuing conflict between the government and the oil companies, beginning with the Mexican revo-
lution of 1910, Mexico finally nationalized and expropriated
the foreign oil companies in 1938. The reasons given for
this action were: inadequate conservation of existing reserves, a lack of interest on the part of the companies in
exploring for new reserves, and unfair labor practices.
The oil companies had refused to accept a wage increase
awarded by the Mexican Federal Labor Board, which was higher
than that demanded by the striking oil workers. In 1938 a
national oil company, Pemex (Petroleos Mexicanos), was co-
tablished with responsibility for all oil operations from
exploration to final sales.[2] The Mexicans consider the
nationalization of Mexico's oil a sign of real economic independence and are very sensitive about it. The anniversary
of the oil nationalization is celebrated every year with
special ceremonies. Mexico's nationalization of its oil was
among the principal factors stimulating the oil companies'
accelerated exploitation and exploration efforts in Venezuela.

Mexico's oil exports drastically declined with the
nationalization in 1938. The international oil companies
applied a very effective oil boycott to Mexico's oil exports,
as well as on its imports of goods and services to run the
oil industry, through their control of oil markets and
threats of legal action.[3] Peter Odell estimated that Mexico
lost about $25 million a year in foreign exchange as a re-
sult of inability to export oil.[4] When the conflict with
the companies was settled in 1945, Mexico was free to export
oil again. However, Pemex was not able to do so as

---

*Barrels per day.

production was barely keeping up with domestic consumption which was steadily increasing.

Like many nationalized industries, Pemex for many years had to cope with government interference in its pricing policies. The government had adopted the policy of providing increasing energy supplies to the domestic customers at low prices. For a long time all domestically sold oil product prices were frozen and low prices led to soaring demand. As operating costs greatly increased, Pemex could not obtain the profits needed for further investment in exploration work, or even the cash needed to fully exploit the new reserves when they were found. In addition, when Pemex was in a position to export some oil, the government then in power decided that the country should economically benefit from exporting refined products instead of crude oil. As there were not many buyers for Mexico's refined products, Mexico lost an estimated $50 million a year in foreign exchange that could have come from oil exports. Pemex had to borrow funds from foreign sources to increase exploration and production.[5]

In the early 1970s Mexico became a net importer of oil to meet its domestic consumption requirements. However, when Mexico's balance of payments began to be adversely affected by the imported oil bills, in early 1974, the prices of domestically sold oil products had to be virtually doubled. Thus, it took Pemex nearly 25 years to obtain the right to act more like a commercial enterprise and charge reasonable prices for its oil products. Consequently, Pemex was able to achieve financial respectability and contract long-term foreign and private loans at favorable rates of interest.[6] Increased prices for domestic oil sales also enabled Pemex to devote more funds for oil exploration and production. This, combined with the discovery of new oil reserves near Tampico and Tuxpan during 1968-70 and especially in the Chiapas-Tabasco area in 1972, again made Mexico self-suffucient in oil by 1974. In the fall of 1974, Mexico resumed exporting 35,000 b/d of crude oil at a price of $11.50 a barrel. Exports of refined products, which were valued at #35 million in 1973, also continued.

I

OIL, NATURAL GAS LIQUIDS, AND GAS RESERVES

As Pemex has until recently been rather conservative and secretive in its estimates of Mexico's hydrocarbon reserves, these reserves have been the subject of considerable speculation. Figures that appeared in the *Oil and Gas Journal* for the years 1968-1978 concerning these reserves as shown in Table 1 indicate the extent and variability of this speculation.

Table 1

MEXICO'S ESTIMATED OIL RESERVES
(in billions of barrels)

1968	1970	1972	1973	1974	1975	1976	1977	1978
5.5	3.2	4.5	2.8	3.6	13.6	9.5+	7.0+	14.0+

+ Revised
SOURCE: *The Oil and Gas Journal.*

Pemex's estimates of Mexico oil reserves include oil, natural gas liquids, and natural gas under the general heading of "hydrocarbon reserves." The potential reserve figures include both proved and probable reserves as well as the potential reserves. Pemex's proved oil reserves include only the wells in production.

According to Tanzer, Pemex invested an estimated $600 million in exploration between 1938 and 1966 and found 7 billion barrels of hydrocarbon reserves.[7]

Some prolific fields were found such as Poza Rica in the Central Zone discovered in 1930, which is 7,090 feet deep and is still producing 47,187 b/d; San Andres in the Central Zone, discovered in 1956, is 10,410 feet deep and still producing 29,223 b/d; Ogarrio in the Southern Zone, discovered in 1957, is 5,790 feet deep and still producing 14,259 b/d.

Between 1966 and 1969 Pemex made some new discoveries—and between 1969-71 substantial discoveries—especially in the Gulf of Mexico off Tampico and Tuxpan. Among these, Arenque, discovered in 1970 in the North Zone off Tampico, was drilled at a depth of 11,362 feet and was producing 3,243 b/d in 1971 and is currently producing about 20,858 b/d. The

93

Table 2

DISCOVERY DATE AND DAILY AVERAGE PRODUCTION
OF MAJOR MEXICAN OIL FIELDS* FOR
FIRST SIX MONTHS OF THE YEAR
(in barrels per day)

Name of Field and Discovery Date	1971	1972	1973	1974	1975	1976	1977
Poza Rica, 1930	52,771	58,009	56,912	55,874	55,783	52,994	47,187
Cuichapa, 1935	24,369	44,925	39,341	30,090	29,020	23,068	18,794
San Andres, 1956	36,581	35,045	32,845	30,634	30,857	28,832	29,223
Ogarrio, 1957	9,957	11,634	15,393	17,763	22,995	19,538	14,259
Cinco Presidentes, 1960	43,312	38,442	33,028	27,444	21,378	18,214	17,625
Arenque, 1970	3,243	7,305	7,500	22,286	19,458	16,105	20,858
Cactus, 1972	--	--	7,930	30,474	60,390	79,519	87,142
Sitio Grande, 1972	--	--	3,178	65,979	89,853	42,124	42,262
Bagre, 1973	--	--	--	5,307	15,730	17,160	11,945
Samaria (Cret.), 1973	--	--	--	12,160	105,318	223,700	287,515
Cunduacán, 1974	--	--	--	--	18,594	65,601	141,468
Iride, 1974	--	--	--	--	7,560	8,379	15,171
Nispero, 1974	--	--	--	--	3,315	5,545	20,788
Other**	253,088	246,376	242,127	215,537	200,515	188,166	173,469
TOTAL	423,321	441,736	438,405	513,548	680,766	788,945	927,606

* Producing over 10,000 b/d in 1977.
** Total of all others producing less than 10,000 b/d in 1977.

Source: *The Oil and Gas Journal.*

94

discoveries made in the Marine Golden Lane, off Tuxpan, were
also quite important.[8] However, the most important discoveries were made in the Reforma area of the Chiapas-Tabasco
region between 1972 and 1974. The majority of the existing
wells were, until then, located in a narrow strip along the
Gulf of Mexico.

In 1972, two important discoveries were made in the
previously unproductive state of Chiapas. One of these,
Sitio Grande, was drilled at 13,766 feet and tested at 1,720
b/d, was producing 3,178 b/d in 1973 and is currently producing 42,162 b/d. The other, Cactus 1, was drilled at 12,333
feet, tested 2,500 b/d, was producing 7,930 b/d in 1973 and
is currently producing 87,142 b/d. The more recent wells
have been drilled at much greater depths than ever done before in Mexico.

While all the previous discoveries were associated with
salt domes, the new discoveries were Cretaceous.[9]

In 1973, the Samaria field was discovered in the same
area called Reforma, which was named after the Mexican revolu-
tion. Samaria was drilled at 14,209 feet, was producing
12,160 b/d in 1974 and is currently producing 287,515 b/d.
In 1973, Bagre was also discovered. It was drilled at 10,919
feet, was producing 5,307 b/d in 1974 and is currently producing 11,945 b/d.

The production of Mexican oil fields compares favorably
with the average fields in the Middle East; however, it
cannot be compared with the production of some of the more
prolific Middle Eastern fields. For example, in Saudi Arabia
Ghawar produces 5.3 million b/d and Safaniya 1.4 million b/d;
in Iran Marun produces 1.4 million b/d and Ahwaz Asmari 1.0
million b/d; in Kuwait Burgan also produces 1.0 million b/d.
For the production of major Mexican fields, see Table 2.

Pemex launched a $3 billion, 3-year development plan
for 1974-1976, which was the largest investment for hydrocarbon exploration and development in Mexico's history. Of
this $3 billion, $240 million was earmarked for geological
and seismic studies, $480 million for drilling 730 wildcats
and $248 million for developmental drillings. Pemex esti-
mated that it would drill 1,600 wells in this period. This
plan called for oil and gas exploration in 21 largely unexplored states as well as in new offshore tracts in both the
Gulf of Mexico and on the Pacific Ocean continental shelf.[10]

In 1974, Nispero (12,993 feet deep, was producing 3,315
b/d in 1975 and is currently producing 20,778 b/d), Cunduacán
(13,442 feet deep, was producing 18,594 b/d in 1975 and is
currently producing 141,468 b/d), Marsopa (10,198 feet deep,
was producing 5,800 b/d in 1976 and is currently producing
5,114 b/d), and Iride (14,596 feet deep--the deepest to date--
was producing 7,560 b/d in 1975 and is currently producing
15,171 b/d) were discovered. In spite of these prolific
discoveries, Pemex's estimate of Mexico's proved hydrocarbon

reserves at the end of 1974 was 5.8 billion barrels. Pemex
had been following a conservative and secretive attitude
concerning Mexico's probable reserves, as sufficient evidence was not available to substantiate claims of larger
reserves.

The discovery of another major oil field in southeastern
Mexico in the area of Cotaxtla, Veracruz, was made in March
1975. At that time, eight wells were already in operation
at Cotaxtla producing 4,000 b/d of crude oil and 10 million
cfd* of natural gas.[11] In June 1975, Mexico's reserve estimates were raised to 20 billion proved and probable reserves.[12]
At the end of December 1976, Pemex gave an estimate of 11.1
billion barrels of proved and more than 60 billion barrels
of probable hydrocarbon reserves. In 1976, 336 wells were
completed and 162 rigs were in use. Drilling was very active
in the Reforma area as well as in Poza Rica, Papaloapan,
Ebano and in the gas-rich northeastern region.[13] By the end
of 1976, only about 10% of the country's oil-bearing terri-
tory--some 2.5 million square kilometers onshore and offshore
--had been explored. The *Wall Street Journal* of October 26,
1977 reported that during the 39 years since the oil nationalization, Pemex had drilled only 13,000 wells as opposed to
40,000 wells drilled in the United States in 1976 alone.

However, in spite of monumental natural and operational
obstacles, such as heavy rainfall, existence of treacherous
swamps, and the necessity of very deep drilling, progress in
the Reforma area has been amazingly fast.[14]

In 1977, Pemex adopted an ambitious $15.5 billion** de-
velopment program which aims at reaching an oil and natural
gas liquids production level of 2.2 million b/d, a gas production level of 2.2 million b/d, a gas production level of
4 billion cfd, an export level of 1.1 million b/d of oil; a
refining capacity of 1.7 million b/d; and a petrochemical
output of 15.5 million tons by 1982. According to the Director General of Pemex, Jorge Diaz Serrano, the company's
philosophy concerning oil exploration is to concentrate its
efforts in the areas which offer the best probabilities
without rejecting the less promising areas.[15]

The 1977 development program calls for the surveying of
1.2 million square kilometers of prospective sedimentary
areas and the drilling of 1,324 wildcats and 2,152 development wells. As 418 wells were planned for 1977, of which 98
were to be exploratory wells, Pemex will have to drill at
least 612 wells a year beginning in 1978. Of the $15.5 bil-
lion, $7.3 billion will be used for drilling and production
development programs and $1.2 billion for exploration. Of
the planned wells, 21 exploratory wells will be in Reforma.
Deep drilling, which may run as deep as 19,000 feet, will be

---

* Cubic feet per day.
**Later increased to $17.0 billion.

required and more rigs will be needed. Pemex has about 50
rigs capable of drilling to 20,000 feet. 447 of the planned
development wells will be in Tabasco and Chiapas [16]
     Offshore exploration and development in the Campeche
Sound will constitute a substantial portion of the 6-year
plan. Pemex contracted for five offshore rigs in 1977 and
two in early 1978 to work in this area. In 1977 and 1978,
24 exploration wells were to be drilled in the Campeche
Sound. 120 development wells were allocated to the area.
Among the wells tested in this area, Chac 1 flowed at about
3,000 b/d in 1976, and in the spring of 1977 Bacab 1 flowed
at about 2,000 b/d. According to Pemex's estimates, by 1983
300,000 b/d will be produced from the five wells it plans to
develop in this region. According to the latest interpretation of seismic data from detailed offshore and onshore
surveys, the Sound of Campeche is not a prolongation of the
Reforma fields but part of a huge new structural trend. It
is estimated that this new oil area extends all the way from
south of Ciudad Pemex to at least 214 miles off Ciudad del
Carmen Island. Although it has similar characteristics with
those of the Reforma area, it has much larger potential in
calcareous tertiary formations.*
     Pemex also contracted for one offshore rig in 1977 to
work the Sebastian Vizcaino Bay on the Pacific Coast to test
the offshore potential of that region.[17]
     In 1977, six new onshore wells--Artesa, Paredon,
Giraldas, Copano, Sunuapa and Tepotzingo--were discovered
in the Reforma area, bringing the number of successfully
tested structures in this region to 23. There were still
70 to be tested. It was also discovered in 1977 that the
above-mentioned Samaria, Iride and two smaller fields were a
single accumulation with 2.5 billion barrels of reserves from
which most of the Reforma production comes. This field is
now called Bermudez.[18]
     In June 1977 Pemex estimated the proved hydrocarbon
reserves of Mexico at 14 billion barrels (an increase of
27%) and the probable reserves at 2.8 billion barrels,[19]
in keeping with the recently adopted policy of evaluating
Mexico's hydrocarbon reserves every six months. At the end
of December 1977, Pemex raised the estimates of proved reserves to 16.8 billion, probable to 30 billion, and potential

---

*According to *Webster's New World Dictionary*, calcareous refers to things of, like, or containing calcium carbonate,
 calcium, or lime.

(or possible) to 120 billion barrels.* About 65% or 10.4 billion barrels of the total proved reserves are estimated to be oil and natural gas liquids and about 35%, or the equivalent of 5.6 billion barrels of oil (31 trillion cubic feet), to be natural gas reserves. Of the 30 billion probable reserves, 65% or 20 billion barrels of the total are estimated to be oil and natural gas liquids and 35%, or the equivalent of 10 billion barrels of oil (nearly 61 trillion cubic feet) to be natural gas.[20]

If these estimates are correct, Mexico's reserves might compare favorably with those of some of the major oil producing countries. (For a comparison of Mexico's proved oil reserves with those of OPEC and other major oil producers of the world see: Table 3.) A Pemex official claimed that the revised reserve estimates resulted from seismic surveys and intensive exploratory drilling in the Reforma area.

The true extent of Mexico's potential reserves can only be determined by extensive drilling which will require many years and considerable funds.

---

*As of January 2, 1979, Pemex estimated that Mexico's proved hydrocarbon reserves had increased to 40.1 billion, probable to 44.6 and potential to 200 billion barrels as a result of big finds in the Chincontepec Basin in the Tampico area and the Campeche region in the Yucatan Peninsula. As the total includes 65% crude oil and natural gas liquids and 35% natural gas, the proved crude oil and natural gas liquids reserves would be about 26 billion barrels.

Table 3

ESTIMATED PROVED OIL RESERVES OF SELECTED COUNTRIES
(in billions of barrels)

COUNTRY	1970	1973	1976	1977	1978	1979	Percent of World Total 1979
Saudi Arabia	128.5	138.0	148.6	151.4	150.0	166.0	25.8
Kuwait	67.1	64.9	68.0	67.4	67.0	66.2	
Iran	70.0	65.0	64.5	63.0[a]	62.0	59.0	
Iraq	32.0	29.0	34.3	34.0	34.5	32.1	
UEA	11.8	23.0	32.2	31.2	32.4	31.3	
Libya	29.2	30.4	26.1	25.8	25.0	24.3	
Nigeria	9.3	15.0	20.2	19.5	18.7	18.2	
Venezuela	14.0	13.7	17.7	15.3	18.2	19.0	
Indonesia	10.0	10.0	14.0	10.5	10.0	10.2	
Algeria	n.a.	n.a.	7.4	6.8	6.6	6.3	
Neutral Zone	25.7	16.0	6.4	6.3	6.2	6.5	
Qatar	4.3	6.5	5.9	5.7	5.6	4.0	
Gabon	0.7	1.5	2.2	2.1	2.1	2.0	
Ecuador	0.6	5.7	2.5	1.7	1.6	1.2	
TOTAL OPEC			449.9[b]	440.4[b]	440.0[b]	444.9[b]	69.3
Non-OPEC:							
USSR	77.0	75.0	80.4	78.1	75.0	71.0	11.0
USA	37.1	36.8	33.0	31.3	30.0	28.5	4.4
W. Europe	3.7	8.6	25.1	24.1	26.9	29.0	3.7
PRC	20.0	19.5	20.0	20.0	20.0	20.0	3.1
Mexico	3.2	3.0	9.5	7.0	14.0	16.0[c]	2.4
Canada	10.7	10.7	7.1	6.2	6.0	6.0	0.9
Others	--	--	34.2	33.2	34.6	31.2	4.8
WORLD TOTAL			658.7[b]	640.4[b]	645.9[b]	641.6[b]	

[a]National Iranian Oil Company later revised this figure to 90 billion barrels.
[b]Details may not add to totals due to rounding.
[c]According to the latest estimates by Pemex, Mexico's proved hydrocarbon
reserves as of January 1, 1979, are 40.1 billion barrels. About 65 percent
or 26 billion barrels of this are estimated to be crude oil and natural
gas liquids.

SOURCE: *The Oil and Gas Journal.*

II

CURRENT OIL, NATURAL GAS LIQUIDS, AND NATURAL GAS PRODUCTION

Oil and Natural Gas Liquids

Mexico's oil and natural gas liquids production has been increasing steadily during the last decade (See Table 4). Production passed the 1 million b/d mark for the first time in February 1977.[21] Production in January 1978 was 1.2 million b/d and reached an average of 1.3 million b/d during 1978. This brought Mexico's oil production to the level of several major oil producing countries (See Table 5). As already mentioned, Pemex was aiming to reach a production level of 2.2 million b/d by 1982; however, it was reported in March 1978 that Pemex now expects to reach this level by 1980.* The Central Intelligence Agency also estimates that this level will be reached by that date. According to CIA projections, Mexico's 1985 production would range between 3.0 and 4.0 million b/d, depending on development policies, but would not be able to exceed 5 to 6 million b/d by that date.[22]

Table 4

MEXICO'S ESTIMATED OIL PRODUCTION
(in thousand barrels per day)

1968	1970	1973	1974	1975	1976	1977
383.5	427.4	478.0	513.5	710.0	850.0	990.0

SOURCE: *The Oil and Gas Journal*

---

*In May 1979, Secretary of National Properties and Industrial Development, José Andrés de Oteyza said that Mexico was planning to produce an additional quarter of a million b/d of crude oil above the 2.25 million b/d originally projected for the end of 1980.

Table 5

ESTIMATED OIL PRODUCTION OF MAJOR PRODUCERS
(in millions of barrels per day)

COUNTRY	1970	1973	1976	1977	1978	Percent of World Total 1978
Saudi Arabia	3.4	7.4	8.6	9.0	7.8	13.0
Iran	3.7	6.0	5.9	5.7	5.3	
Venezuela	3.6	3.4	2.3	2.3	2.2	
Iraq	1.5	1.9	2.0	2.2	2.5	
Nigeria	1.0	2.0	2.0	2.2	1.8	
Libya	1.0	2.1	1.9	2.0	2.1	
UAE	0.7	1.5	1.9	1.8	1.8	
Kuwait	2.7	2.9	1.8	1.7	1.9	
Indonesia	0.9	1.3	1.5	1.7	1.7	
Algeria	1.0	1.0	1.0	1.0	1.3	
Neutral Zone	0.5	0.5	0.5	0.4	0.4	
Qatar	0.4	0.6	0.5	0.4	0.5	
Ecuador	negl.	0.2	0.2	0.2	0.2	
Gabon	0.1	0.1	0.2	0.2	0.2	
TOTAL OPEC			30.9[a]	31.0[a]	29.5[a]	49.1
Non-OPEC:						
USSR	7.1	9.2	10.5	11.0	11.4	19.0
USA	9.5	9.2	8.1	8.2	8.7	14.4
PRC	0.4	1.0	1.7	1.8	2.0	0.3
W. Europe	0.4	0.3	0.8	1.3	1.8	0.3
Canada	1.2	1.8	1.3	1.4	1.3	0.2
Mexico	0.4	0.5	0.9	1.0	1.3	0.2
Others	2.7	3.3	3.5	3.7	4.1	6.8
WORLD TOTAL	44.9[a]	55.2[a]	57.2[a]	59.5[a]	60.0[a]	

[a]Details may not add to totals due to rounding.

SOURCE: *The Oil and Gas Journal.*

In accordance with the six-year plan (1976-1982) adopted by Pemex, the production base will be expanded by the discovery and development of new fields in established areas such as Chiapas, Tabasco, Cotaxla, Nuevo Laredo, Sebastian Vizcaino, and the Campeche Sound. The Central Plateau, Coahuila, Chihuahua, and the Chiapas Mountains also have long-term prospects. Intensified primary exploration of existing fields, secondary recovery and the development of new off-shore reserves will be the main components of the production program.[23]

Natural Gas

In 1977 Mexico's proved and probable gas reserves in the Reforma Tabasco-Chiapas area were estimated to be 9.66 trillion cubic feet (much lower than the estimates on p. 8) by a six-company U.S. group that was planning to import Mexican gas into the U.S. Potential additional reserves were estimated at 20.65 trillion cubic feet.[24]

Most of the gas production in Mexico is associated with crude oil production, although there also are natural gas fields independent of oil fields. Recently, Mexico's oil fields have been producing much higher ratios of gas per barrel of oil than was previously estimated. Some of the newer fields have been producing about 6,000 cfd of gas per barrel of oil--which is exceptionally high--as against a ratio of 1:1000 for older wells.[25] Mexico's gas production in June 1977 was 2,030 million cfd. It was expected to reach 2,560 million cfd by the end of 1978. According to the six-year plan (1976-1982), the projected gas production in 1982 is estimated to reach 4 billion cfd. However, as gas production ratios and oil yields have been substantially higher in the recent discoveries, it is thought that this projection for 1982 is very conservative. Some Pemex engineers believe that by 1982 Reforma may be producing 8-10 billion cfd. If 4 billion cfd of this could be absorbed by the local markets and 2 billion cfd exported, there would still remain a surplus of 2 billion cfd.[26]

Mexico has launched an ambitious program for gas treatment plants to be able to handle the large amounts of gas production from the oil fields that are being developed. Most of the gas is being used or will be used in Pemex's petrochemical complexes in southern and central Mexico and a proportion of it is being fed into Mexico's gas system. As it will be discussed later, Pemex was also planning to export some of this gas by means of extending a pipeline, that it plans to build to the industrial centers in northern Mexico, to the U.S. border and linking it to the existing pipelines in the U.S. The pipeline to northern Mexico is planned to be over 737 miles and 48 inches in diameter with an initial capacity of 1.3 billion cfd and eventually expanding to 2.7 billion cfd.

The contracts for supplying about 737 miles of the pipeline to northern Mexico went to Germany's Mannesmann at
$30.18 million, Japan's Marubeni at $65.06 million, France's
Vallourec at $48.98 million, Mexico's Tubacero at $42.63
million, and Italy's Italsider at $34.23 million. The only
U.S. firm that was to supply part of the pipeline, U.S. Steel
at $33.45 million, had to cancel its contract because it
could not meet the delivery date. That contract also went
to Germany's Mannesmann in addition to the order it had won
earlier. Construction work was to be divided into five sep-
arate sections, with completion by December 1978, to an initial capacity of 1.3 billion cfd, increasing to 2.5 billion
cfd by May 1979. The pipeline was inaugurated by President
López Portillo on March 26, 1979, with an initial capacity
of 800 MM cfd.[27]

III

PLANS FOR NEW REFINERIES
AND PETROCHEMICAL PLANTS

Refineries

Mexico plans to continue refining at least 50 percent of its oil output and thus benefit from the added value itself.

There were nine refineries in Mexico in December 1977 with a crude refining capacity of 850,000 b/d. The largest Mexican refineries are located at Minatitlán (270,000 b/d), Cadereyta (235,000 b/d), Salamanca (210,000 b/d.), and Ciudad Madero (185,000 b/d).[28]

Pemex plans to increase the country's refining capacity to 1.7 million b/d by 1982. It aims to "insure the country's self-sufficiency in refined products through 1982 and, at the same time, have excess processing capacity to maximize product exports."[29] In order to achieve its aims, Pemex intends to improve the existing facilities and to develop new refining centers at Salina Cruz to serve the Pacific Coast and at Cadereyta to serve the north and the northeast. New refining units are also being constructed in Ciudad Madero, Minatitlán, and Salamanca.[30]

By 1977, pipelines were built to carry Reforma oil to every single refinery both existing and under construction. Eventually, all Mexican refineries will at least be partially receiving Reforma oil.

Petrochemical Plants

Pemex also has ambitious plans for tripling its petrochemical output by 1982. Its aim is to achieve self-sufficiency in basic petrochemicals by 1979 and subsequently develop large volumes of feedstocks for export. As already mentioned, Pemex hopes to optimize the utilization of Reforma associated-gas for petrochemicals.

Pemex expects the production of basic petrochemicals to increase from 4.9 million tons in 1976 to 21.7 million metric tons in 1982.[31] According to Pemex Director General Jorge Diaz Serrano, "the number of basic petrochemical plants will increase from 60 to 115 and the number of products will also increase from 38 to 44."[32]

The dominant products will be ammonia, aromatics, ethylene, and propylene. Much of the construction will be concentrated in two major complexes at La Cangrajera and Allende (both in Vera Cruz). Pemex's ammonia program is expected to make it one of the largest ammonia producers in the world,

with output increasing from 4,000 tons in 1977 to 13,000 in
1982. The ethylene capacity will increase from 240,000 tons
in 1977 to 2 million tons by 1982. The output of two new
products for Mexico will be 180,000 tons a year of polyethyl-
ene from ethylene by 1982 and 300,000 tons a year of propylene by 1980.[33]

IV

FINANCIAL PROSPECTS AND PROBLEMS

In order to reach its production and export goals in
the early 1980s, Mexico's oil industry will have to expand
smoothly and, capital being the major requirement for finding
and developing oil, will need large amounts of investments.
As already mentioned, the six-year (1976-1982) development
plan for the oil and gas sector requires an investment of
$15.5 billion. Pemex expects to finance this program on a
one-for-one basis--one dollar of Mexican capital for every
dollar borrowed outside.[34] European and U.S. bankers have
considered Mexico a blue chip risk as a borrower, mainly
because of its record of political stability and now consider
it even more so because of its oil wealth.

For example, Mexico borrowed a $350 million Eurocredit
for Pemex in the spring of 1977[35] and in October 1977, Pemex
closed a $90 million private placement with 12 U.S. institu-
tional investors led by Prudential Insurance Company of America. The issue is for 15 years and carries a fixed interest
rate of 9-7/8 percent. The proceeds were to be used for the
company's 1977 investment program.[36] In November 1977,
Mexico borrowed a $1.2 billion Eurodollar credit financed by
an international consortium of 74 banks. The loan will be
earmarked for the exploration and exploitation of oil, the
production of nuclear and hydroelectric energy, and modernization of agriculture.[37] In December 1977, Pemex arranged a
credit line valued at the equivalent of $200 million with a
group of French banks for the purchase of offshore drilling
and gas production equipment, pipelines and other oil and
gas production equipment from French companies. Two specific
agreements were already signed with two French steel concerns,
Vallourec and Usinor.[38]

By the end of 1977, Mexico's total foreign debt was expected to reach $30 billion. Mexico's gross debt incurred
in 1977 was $7.5 billion, including the $1.2 billion loan
borrowed by the government of Mexico. According to the
terms agreed with the International Monetary Fund, of this
$7.5 billion, Mexico was to pay $4.5 billion to service its
foreign debt in 1977, reducing its net borrowing for that
year to $3 billion. The ceiling agreed upon with the IMF for
Mexico's 1978 new borrowing abroad is $2 billion.[39]

In January 1978, Pemex obtained a $104 million credit
from a syndicate of 16 Japanese banks and two insurance companies to buy equipment for oil exploration and refining and
petrochemical plants.[40]

In early 1978, Pemex was negotiating to borrow a medium-term loan of $800 million--which was later increased to $1 billion--from 15 international banks for its oil exploration, development and production program.[41] Francis Ghiles in *Financial Times* of February 21, 1978, noted that "Every major loan for a Mexican borrower, be it state owned or state guaranteed, has in the past 10 months marked an improvement in terms compared with the last one." This specific loan of $1 billion would be for ten years at an interest rate 1-1/4 percentage points above the London interbank rate. Some observers wonder if this extensive borrowing and expansion of refineries and petrochemical plants is wise for Mexico in the long run.[42]

Domestically Pemex was planning to issue about $90 million worth of "petrobonds" in early 1977 that would have a maturity value pegged to the world price of oil. The 7 percent "petrobonds," each to have a nominal value of about $45, were to be payable in three years and were to be backed by 6.5 million barrels of oil acquired from Pemex by Nacional Financiera, the government development bank. The purchase of the bonds by individuals was to be limited to a maximum of $45,000 and by institutions to a maximum of $225,000.[43]

Mexico has no intention, for the time being, of inviting the foreign oil companies back to accelerate its exploration, production and other oil and gas activities. It is generally agreed that the level of technical competence at Pemex is quite high. It has its own geologists, petroleum engineers, chemists, drilling crews and other technical personnel. However, Leonard Silk of *The New York Times* thinks that this resistance against bringing in foreign oil companies may change in time "if the Mexican government gets in too far over its head in its foreign borrowing and does not control its domestic inflation."[44]

Pemex hires some foreign firms, including American companies, on a service-contract or consulting basis for some jobs such as plant and pipeline construction and offshore drilling, but they are expected to keep a low profile.[45]

Pemex is a fully integrated national oil company and it is big. For example, in 1975 it employed 80,000 people, more than any other business in Mexico. It occupies the biggest building in Mexico and its direct taxes to the government, which accounted for 12 percent of the money received by the Mexican treasury in 1975, are more than the total paid by all private industry.[46]

Reportedly there are problems of corruption, nepotism and featherbedding and a low level of worker productivity in Pemex. The Mexican oil workers union is a labor organization that enjoys some unusual powers and privileges dating back to the 1938 nationalization. For example, Pemex had to grant the union the power to bid for Pemex contracts and, consequently, the union operates some businesses on the side.[47]

V

MEXICO'S OIL PRICES AND
ATTITUDES TOWARD OPEC`

Although Mexico's attitudes toward membership in OPEC varied from one period to another, it has not joined OPEC. Nevertheless, the price of its oil has kept up with, and even surpassed, the average OPEC price since Mexico started to export again in 1974.

In October 1974, *The Washington Post* published a statement by a U.S. energy official which said that if Mexico exported the oil found in Chiapas, Tabasco and Veracruz, it could cancel out boycott threats by members of OPEC. Mexico's National Property Minister, Horacio Flores de la Pena, responded to that by saying that the government disclaimed an "interpretation of that type." He added that, "As a pioneering country in the national and absolute control of oil resources, Mexico will never be the Trojan horse of the international (oil) companies." He also said that Mexico would never "favor with its oil resources the economies of other countries to the detriment of its own, or to the detriment of the third world." He stated that Mexico would not automatically sell its extra output to U.S. companies but would try to obtain the highest world price. According to him, Mexico had no objections to eventually joining OPEC but its foreseeable exports in the next few years would not be large enough to justify immediate membership. He stressed that Mexico would take care that its exports of crude "in no way" helped "depress the price of the product in the world market" and would take care to conform to OPEC price structures. Flores de la Pena later said that Mexico would try as soon as possible to apply for observer status in OPEC.[48]

When former President Echeverria met with former President Ford in October 1974 he said, during a joint press conference, that the U.S. would have to pay the world market price if it wants to buy Mexican oil.[49]

It was reported by the *Wall Street Journal* of February 11, 1975 that Mexico had quietly decided to stay out of OPEC while at the same time refusing to accept less than the high oil prices established by OPEC. In that way Mexico would have the best of both worlds. It would get the benefit of high oil prices but would not be tied to OPEC policies. It was also thought that the new U.S. trade reform act that denied trade preferences to OPEC member countries made membership in OPEC less attractive to Mexico.

However, President Echeverria announced at a press conference in Cancun, Mexico in May 1975, just before the Shah

of Iran concluded a five-day visit there, that Mexico would
join OPEC if invited.[50] On the other hand, Mexico's President
-elect, Jose Lopez Portillo, stated in September 1976 that
Mexico's oil exports were not sufficient to justify OPEC membership, but it would nevertheless always sell oil at "international market prices."[51] In August 1976, Mexico's crude
export price was $12.15 per barrel whereas the price for
Arabian light-34 (the OPEC "marker crude") was $11.51, the
highest Algerian crude price $13.05 and the highest Nigerian
crude price $13.19.[52] The basic sales price for Mexican
crude was increased to $13.35 per barrel in February 1977 and
to $13.40 in August 1977.

Platt's Oilgram News reported on December 2, 1977 that
President Jose Lopez Portillo, when asked if Mexico would follow any price increase voted by OPEC at its next meeting at
Caracas, responded by saying, "If there is a commercial system
that fixes prices, we will follow it. We cannot be scabs of
a system that protects fair valuation of a raw material, nor
can we deprive the country of the benefits of a commercial
price available for our products. We subsidize nobody."

The current Director General of Pemex, Jorge Diaz
Serrano, also reiterated that Mexico does not intend to join
OPEC. He said that they do not "believe in joining cartels...
The minute you join such an organization you lose the free
determination of your actions, and that is the essence of the
government policy." He added that "OPEC was created in the
1960s for the purpose of protecting the nation's interests
from the companies that worked in their territories. We
solved that problem in 1938 when the oil companies were
nationalized."[53]

For the third quarter of 1978, the price of Mexico's
34-gravity Isthmus crude was reduced to $13.10 a barrel, and
on December 19, 1978, increased by 4.6 percent to $14.10
a barrel as of January 1, 1979. In April 1979, Mexico increased the price of its crude oil by $3 to $17.10 a barrel--
a 21 percent increase. Again in July 1979, it raised the
price of its crude by 32.2 percent to $22.60 to remain in
effect until September 30, 1979.

VI

MEXICO'S CRUDE OIL, REFINED PRODUCTS AND
GAS EXPORT POTENTIAL AND PLANS

As already mentioned, Mexico exported relatively small amounts of crude oil and refined products after 1938. Export amounts fluctuated between 1938 and the early 1970s.[54] However, in the early 1970s Mexico became a net importer of oil. It resumed exporting in 1974 with 35,000 b/d of crude oil and continued to export some refined oil products.

Mexico's export policies have been very different under the presidencies of Louis Echeverria and Jose Lopez Portillo.

Under President Echeverria, Mexico's policy was to export only as much crude as necessary to strengthen the country's balance of payments over the short term. His goal was to enlarge Mexico's refining capacity and establish a petrochemical industry in order to sell mostly finished products such as gasoline and raw materials for making plastics.

In 1975, Echeverria said that as far as the surplus sales were concerned, Mexico's oil policy would be "profoundly nationalist and anti-imperialist." According to the then Director General of Pemex, Dorali Jaime, Mexico would seek "maximum diversification of purchases, giving special priority to the needs of the developing countries, above all, those of Latin America." An official of Pemex also stated that the country's oil strategy was not "based on massive exports" and that Mexico was not going to follow "Venezuela's example of unrestrained exports." As already mentioned, this attitude was seen as a rebuff to U.S. news reports that Mexico would be willing to export its surplus oil entirely to the U.S. and at a lower price than that charged by OPEC.[55]

On the other hand, President Jose Lopez Portillo sees Mexico's future linked more and more with the United States, because of the greater interdependence between the two countries, than with the "third world." In order to emphasize this point, he authorized emergency shipments of oil and gas to the U.S. before his visit here in February 1977. His approach is not interpreted as a denial of the ties with the "third world" that were stressed by Echeverria, but a shifting of Mexico's international role back toward more traditional linkages. His government sees Mexico's future "undeniably tied with the U.S."[56]

The Lopez Portillo government hopes that during its six-year term Mexico will rival, if not surpass, Venezuela as the leading oil exporting country in the Western Hemisphere.

From September 17, 1974 to the same date in September 1975, Mexico exported 28.3 million barrels of oil and gained

$308 million.[57] In 1975, Mexico was exporting about 110,000
b/d to the U.S., Uruguay, Israel and Cuba.[58] In 1976, exports
increased to about 140,000 b/d, mainly to the United States
which brought in a total of $543.9 million. In 1977, exports
averaged 200,000 b/d and during the first few weeks of 1978,
280,000 b/d. Exports were expected to reach 500,000 b/d by
the end of 1978, but averaged only 365,000 b/d because of
technical problems in production, lags in pipeline construction, and bottlenecks at seaports. Consequently, Pemex
failed to meet some of its export commitments. For example,
in May 1979 deliveries were reduced by about 40% to some
customers. However, Mexico still hopes to export over 1 million b/d by the end of 1980. It was estimated that 400,000
b/d of exports would be crude oil and 700,000 would be refined products.

<u>Exports to the United States</u>

     Because of geographic proximity and economic interdepen-
dence, the United States would be the natural market for
Mexico's exports of oil, and especially of natural gas. Mexico's oil exports to the U.S. have been steadily increasing,
as indicated on Table 6. From the point of view of security
of supplies, importing Mexican oil would be advantageous to
the United States. When President Eisenhower introduced mandatory quotas on both crude oil and oil products in 1959, Mexican and Canadian oil and oil products were excluded from the
restrictions as imports from these countries were not considered at risk, because they were not dependent on ocean
transportation.
     It would equally be advantageous for Mexico to export
its oil and natural gas to the United States in view of the
persistent deficit in its trade balance with this country
(See Table 7).
     Furthermore, the sheer burden of feeding its burgeoning
population in the decades ahead will require Mexico to maintain, and perhaps increase, its imports of agricultural commodities from the U.S. Although Mexico exports some fruits,
vegetables and livestock to the U.S. it is and will continue
to be heavily dependent upon imports of U.S. oilseeds, feed
grains and food grains. Mexico's population is now estimated
to be about 65 million, and is growing at a rate of some 3.3
percent a year--in spite of an active population planning program. At this rate of growth, Mexico's population could reach
150 million in less than 25 years. Even if Mexico succeeds
in maintaining its current respectable 1.8 percent per year
increase in agricultural production, the food gap will continue to increase. Prospects for significantly increasing
agricultural growth rates are not high, given the limitations
on land and water, the existing patterns of land-holdings, and
the political resistance to change. In any event, the economics of comparative advantage strongly favor the exchange

112

Table 6

U.S. DIRECT AND INDIRECT[a] OIL IMPORTS BY SOURCE
(in thousands of barrels/day)

COUNTRY	1973	1974	1975	1976	1977	1978[b]
Algeria	151.2	207.1	288.2	438.3	563.1	679.6
Indonesia	237.7	310.9	437.7	569.4	585.0	582.7
Iran	433.7	731.0	524.8	546.5	828.1	914.3
Libya	308.3	40.3	329.3	529.3	848.5	709.7
Nigeria	607.9	912.2	837.8	1,119.2	1,240.8	923.0
Saudi Arabia	740.3	675.2	891.6	1,365.8	1,585.1	1,213.9
UAE	83.6	87.8	154.2	323.2	431.9	464.2
Venezuela	1,633.7	1,457.8	1,030.1	972.2	908.0	852.5
Other OPEC[c]	194.5	217.0	259.3	216.0	388.8	291.9
TOTAL OPEC	4,390.0	4,669.3	4,753.0	6,079.9	7,379.3	6,631.8
Non-OPEC:						
Canada	1,312.9	1,067.6	845.2	599.3	515.5	445.0
Mexico	15.2	8.4	71.4	87.1	179.3	255.4
Others	2,469.6	2,160.6	1,824.9	1,921.1	2,559.5	2,447.3
TOTAL	7,664.2	7,521.7	7,188.4	8,313.9	9,972.9	9,136.8

[a]Indirect imports refer to U.S. imports of petroleum products, primarily from Caribbean and European areas, that have been refined from crude oil produced in other areas. U.S. imports of these products have been prorated to each OPEC country of origin based on the share of total crude oil supply in the Caribbean and European areas which was imported from each OPEC country.
[b]First seven months.
[c]Includes Ecuador, Gabon, Iraq, Kuwait and Qatar.

SOURCE: U.S. Department of Energy, Energy Information Administration, *Monthly Energy Review* (Washington, D.C.: October 1978), pp. 18-19.

Table 7

MEXICO'S BALANCE OF TRADE WITH THE UNITED STATES
(in millions of U.S. dollars)

	1972	1973	1974	1975	1976	1977
Imports from the U.S.	1,557.5	2,037.5	3,778.6	4,113.2	3,773.2	3,493.1
Exports to the U.S.	1,118.3	1,318.1	1,703.3	1,667.5	1,875.3	2,426.4
Deficit	439.2	719.4	2,075.3	2,445.7	1,897.9	1,066.7

SOURCE: *IMF Directions of Trade Annual,* 1970-1977.

of Mexican oil for U.S. agricultural commodities, rather than
the development by Mexico of self-sufficiency in food
production.

Pemex had agreed to extend to Reynosa (at the U.S. border)
the over 700-mile pipeline it planned to build to carry natural gas from the Reforma fields to the industrial centers of
northern Mexico (See page 12). The pipeline, which then would
have been over 800 miles long, would have been linked to the
existing U.S. pipelines. The line was to have branched at
San Fernando, Vera Cruz, with another line carrying the gas
to hook into already existing Reynosa-Monterrey lines. Completion of the construction was to be in 1979, with an initial
capacity of 1 billion cfd to be increased to 2 billion cfd.
A letter of intent was signed in the summer of 1977 with six
U.S. companies, specially selected by Pemex to ensure the
widest possible distribution of gas in the U.S.--to the east,
south to north, plus California and possibly Chicago. The
U.S. companies involved and their shares of gas were to be:
units of Tenneco, Inc., 37.5%; Texas Eastern Transmission
Corp., 27.5%; El Paso Co., 15%; Transco Cos., 10%, and Southern Natural Resources, Inc. and Florida Gas Co., which would
share the remaining 10%. The firm contract or contracts were
to be negotiated by December 31, 1977 and would initially have
been for six years, to be followed by another six-year contract
if the U.S. companies met the "best offer" made to Pemex. The
agreed-on price was to be $2.60 per thousand cubic feet to
escalate with the price of No. 2 home heating fuel in New York
harbor.

Offers to finance the pipeline came from West Germany,
France, Britain, Italy, Japan and the United States. The U.S.
Export-Import Bank tentatively approved $590 million in loans
at 8.5 percent interest to Pemex--$340 million for the plan-
ned pipeline and the remainder for some other projects to
increase its oil and gas output and petrochemical industry.
This loan was to be Eximbank's first to Pemex. An additional
$250 million was to come from a U.S. private banking consor-
tium headed by Bank of America, Chase Manhattan and City Bank.

However, opposition began to build both in the U.S.
Congress and the administration against the price for the gas
imports from Mexico. U.S. officials opposed the price of
Mexican gas which at $2.60 a thousand cubic feet would be
higher than the Canadian gas imported at $2.16 a thousand
cubic feet. There was fear that if the price asked for Mexican gas were to be accepted, the Canadians would also increase
the price of their gas. The U.S. Secretary of Energy, James
R. Schlesinger, stated that the Carter Administration would
not permit the establishment of a price for Mexican gas that
would be above the price paid to the Canadians. The price for
Mexican gas would also, of course, be much higher than the
U.S. domestic gas prices.

Officials also opposed the linking of the price of the
Mexican gas to the price of No. 2 home-heating fuel in New

114

York harbor because of the automatic escalation as world oil
prices increase. The U.S. government would not have any
control over future prices of gas imports from Mexico under
such an arrangement.

To increase pressure on Administration officials and the
U.S. companies to hold down the price of the Mexican gas,
Senator Adlai Stevenson (D. Ill.) introduced a resolution in
November 1977 to hold up the Eximbank loan to Mexico until
the Department of Energy reviewed the price of the Mexican
gas. Most of the U.S. government officials feel that Mexico
would be obliged "sooner or later" to sell its gas to the
United States--which is the only major consumer within reach
of the pipeline.

As a result, the Eximbank loan which was approved on
December 17, 1977 would be disbursed only after the American
companies had executed "binding contracts" with Pemex for the
purchase of the gas.

However, the companies that had signed the letter of
intent were very upset about this setback, because they consider the prices asked for the gas by Pemex to be a fair and
reasonable price. They claimed that the decision would have
an adverse effect on the U.S. gas supplies early in the next
decade.

The Mexican government presently refuses to decrease
the price that it asks for its gas. In response to a statement by U.S. Energy Secretary Schlesinger that Mexico "sooner
or later" would have to sell its gas to the United States,
Mexican officials declared that Mexico was prepared to use
its gas domestically within the Mexican economy by substituting it largely for fuel oil rather than export it at a lower
price. As the delay has incrcased President Lopez Portillo's
domestic problems, the Mexican government decided to go ahead
with construction of the pipeline, but only as far as Monterrey, about 90 miles from the U.S. border. Pemex placed a
$50 million bond issue in Kuwait in January 1978 and expanded
a credit line with the French foreign-trade bank from 300
million to 700 million francs. A $59.3 million Pemex bond
issue was sold in Bahrain in December 1977 and the company
has been dealing with a dozen non-American banks since mid-
1977. Pemex borrowed $50 million from the Bank of Montreal
in September 1977. Pemex was trying to strengthen its position vis-a-vis the United States in future negotiations
concerning the price of its gas by borrowing outside the
United States.

At the end of President Carter's visit of February 12-14,
1979 to Mexico, he and President López Portillo agreed to
start the government-to-government negotiations for the United
States' purchase of the natural gas. The negotiations were
started on April 3, 1979 by a joint Sub-Cabinet Committee
that was established as a result of the agreement between the
two presidents. Although good progress was being made in the
negotiations, the two sides were reported to be far apart on
some issues, especially the price. In the meantime, Mexico's

Ambassador to Canada, Augustin Barrios Gomez said in Ottawa
in May 1979 that Mexico's price for natural gas exports had
increased to $3.20 per thousand cubic feet.

Most observers think that the negotiations would not be
completed before the end of 1979. However, there is a possibility that they might be concluded before President López
Portillo's scheduled visit to the United States during the
summer of 1979.

If agreement can be reached, it is thought that the volume of the initial deliveries will be less than originally
planned.

Mexico's natural gas production in 1978 reached 2.56
billion cfd. Gas flaring was reported to be reduced to 200
million cfd and the goal for the end of 1979 is to reduce it
to less than 100 million cfd. In April 1979, the Director
General of Pemex, Jorge Diaz Serrano said that Pemex would
have a surplus of 800 million cfd by mid-1979 which would be
available for export. It is estimated that by 1982 the
natural gas surplus would be 2.8 billion cfd if Mexico produces 2.2 million b/d of oil by 1980 as had been planned, and
keep to this ceiling until 1982. However, as already mentioned, a high Mexican official told in May 1979 that Mexico
was planning to produce an additional quarter of a million
b/d of oil a day above the 2.25 b/d originally projected for
the end of 1980. In that case, the gas surplus will be even
more than 2.8 billion cfd by 1982.

Some observers doubt that, if agreement is not reached
between Mexico and the United States, Mexico's industry will
have the capacity to absorb the amount of gas involved, which
represents about 4 percent of the current U.S. requirement.
They also believe that while the wells that produce the associated gas are now capped to avoid flaring, the pressure to
meet crude-oil production targets will necessitate removal
of the caps in the near future. They also feel that Mexico
cannot readily sell the gas to other countries since any
exports would require costly liquefaction facilities. The
only realistic market for the gas is that of the U.S., and
Mexico needs the import revenues from it.

The disagreement with Washington is considered to be
particularly embarassing to President López Portillo, as he
has been sharply criticized by the leftists and nationalists
for his original decision to build the $1.5 million natural
gas pipeline to the U.S. border.[58] However, the Mexicans
believe that time is on their side and are prepared to wait
as long as necessary for the U.S. government to accept their
price requirements.

Exports to Other Countries

Although Pemex is very U.S.-oriented, the controversy
over the gas pipeline, as well as criticism by the leftist
economists that Mexico is increasing its vulnerability to

American pressures by concentrating its exports to one market,
has obliged the Mexican government to seek new markets for its
oil exports. President López Portillo said in an interview
that as Mexico increases its supply of crude and processed
products, it should vary its markets, and that Latin America
and Europe are logical markets.

As already mentioned, Pemex has already been selling oil
to Israel and also sold to Uruguay and Cuba. The amounts of
exports to Israel were increased from 20,000 b/d to 30,000 b/d
in 1977. Pemex made an agreement to sell at least 446,000
barrels of crude oil--later increased to 1.35 million barrels--
at $13.45 a barrel, and $11.5 million worth of petrochemicals
to Spain in 1977. In September 1978, a five-year agreement
was signed with Spain's Cepsa to export 24,000 b/d of crude to
Spain. Italy's ENI was interested in processing Mexican crude
in its refineries and selling the products under a joint Pemex
-ENI brand. Mexico also sold oil to Britain and Sweden.[59]

In December 1978, Pemex signed a ten-year agreement with
France's CFP to export 100,000 b/d of crude oil to France
starting in 1980. In April 1979, an initial agreement was
signed with Sweden to export over 70,000 b/d of crude to that
country starting in 1981. In May 1979, Canada and Mexico
signed a ten-year agreement for exports of up to 100,000 b/d
of Mexican crude oil to Canada in exchange for nuclear technology and as much as 3 million tons of coal. A large number
of joint ventures were also being planned between Canadian
and Mexican companies.[60]

Brazil was intending to sign an agreement with Mexico
to buy 100,000 b/d of Mexican crude oil starting in June 1978
in exchange for sale of agricultural products, such as soy-
beans, and of iron ore. The agreement was not signed because
Brazil found Mexico's price of $13.40 per barrel of crude oil
too high, as only small ships could be used to transport
Brazilian exports in exchange for the crude. Mexico does
not yet have deep water ports to handle supertankers and
decided to store its excess oil temporarily on large offshore
floats to facilitate supertankers' loading while deepening
its major ports. It is preparing deep water terminals at
Pajaritos on the Gulf coast and at Salina Cruz on the Pacific
coast to accommodate supertankers to be able to expand its
oil exports.[61]

Japan is also interested in importing Mexican oil, as
it is trying to diversify the sources of its oil imports.
When Kakuei Tanaka was the Prime Minister of Japan, he paid
a state visit to Mexico in September 1974 and expressed interest in buying Mexican oil. President López Portillo also
visited Japan. As already mentioned, the Japanese company
Marubeni, supplied part of the gas pipeline from the Reforma
fields to Monterrey. In order to be able to receive Mexican
oil, the Japanese proposed to build a multimillion dollar
industrial complex at Mexico's Pacific coast and a pipeline
from the eastern oil fields across the country to the Pacific
coast. In order to help finance oil development projects, a

group of Japanese banks, headed by the Bank of Tokyo, signed
an agreement to provide a $125 million five-year loan at a
margin of 0.625% over the London interbank offered rate
(LIBOR) and a second group of Japanese banks headed by the
Industrial Bank of Japan were to lend $100 million over a ten-
year period at a margin of 0.75%.

Japan and Mexico agreed to start negotiations at the
government level for a regular supply of Mexican oil to Japan.
The negotiations were to start at the end of July 1979.[62]

As a result of Mexico's export commitments to other
countries, the U.S. share of Mexican oil exports is expected
to fall to 60 percent of the total exports by the late 1980s,
instead of the current 80 percent. Thus the U.S. would be
receiving about 660,000 b/d, if Mexico's oil export volume
reaches 1.1 to 1.2 b/d in the late 1980s as projected.

The *Financial Times* of April 6, 1978 reported that Pemex's
Director General Diaz Serrano makes it clear that the production and export targets may be altered one way or the other
in light of market sentiment. There is considerable disagreement in Mexico about the amounts of oil it should produce and
export.

CONCLUSION

Mexico's oil and natural gas reserves and production
have dramatically increased during recent years and turned
Mexico, once again, into a major producer and exporter of oil.
But, contrary to the expectations of many, Mexico's oil will
neither be the automatic solution to Mexico's economic problems nor the solution to the energy problems of the United
States.

According to different estimates, Mexico may be earning
between $3.5 billion and $6 billion from exports of crude oil,
refined products and petrochemicals by 1982--provided oil
prices remain high. The existence of a sustained short-term
glut of oil would lower these export revenue estimates significantly. In any case, Mexico is faced with the same
serious social and economic problems faced by many developing
countries. Its population is growing at an alarming rate,
faster than its GNP growth rate, and is not likely to slow
significantly for decades, if at all. Although its per capita
GNP is relatively high compared to many developing countries--
$1,350 in 1976--its income distribution is one of the most
inequitable. Much of its agricultural production is ineffi-
cient, and the social and technical constraints on the massive
changes needed will probably preclude it from achieving its
goal of becoming self-sufficient in foodstuffs. Unemployment
and underemployment rates are very high and the burgeoning
population growth will make it extremely difficult to lower
these rates very much in the next few years. The planned
construction and expansion of refineries and petrochemical
and other industrial plants will not, by themselves, create
the number of jobs required to reduce unemployment to acceptable levels, and will likely increase Mexico's import requirements, thus further worsening its already serious balance of
payments problem. Whether expanding oil and gas production
will help solve, or merely exacerbate Mexico's economic and
social problems depends in large part on how wisely and
efficiently the revenues are used.

To avoid the disadvantages perceived in excessive dependence on the U.S. for an oil export market, Mexico is
trying very hard to find other export customers--even though
the U.S. is the most "natural" market. However, even if all
or most of its oil and natural gas exports go to the U.S.,
the amounts involved will be comparatively small in view of
the projected U.S. consumption requirements. The CIA, in its
International Energy Situation: Outlook to 1985, forecasts
that U.S. oil demand in 1980 will be between 19.3 and 20.7 m.
b/d; and in 1985 will be between 22.5 and 25.6 m.b/d. If this
consumption level is not decreased through conservation and
use of alternative energy sources, it is very likely that more

than half of this consumption will have to be met by imports.
Even if Mexico reaches its goal of 1.1 million b/d of exports
by 1980 (about half of its production goal), and does not sell
most of it to other customers, this quantity will not significantly alter the U.S. import requirements from the Middle
East. However, the CIA also estimates that Mexico might possibly reach a production level of 5 to 6 million b/d by 1985,
depending on development policies (and, presumably, on capital
and other requirements and world economic conditions). If
Mexico's domestic consumption did not also expand to use this
additional supply, the increased production could yield an
exportable surplus of from 4 to 5 million b/d. In this case,
Mexican oil available for export could represent a more significant source of diversification of supply for the U.S.,
though it would still not be a substitute for heavy dependence
on the Middle East. In addition, if an agreement can be
reached on the price, the sale of Mexican natural gas could
also help meet U.S. energy requirements in the 1980s.

FOOTNOTES

1. <u>The Wall Street Journal</u>, October 26, 1977.
2. Peter R. Odell, "The Oil Industry in Latin America," in
   Edith T. Penrose, <u>The Large International Firm in Developing Countries</u> (Cambridge: The MIT Press, 1968),
   pp. 288-289; Michael Tanzer, <u>The Political Economy of
   International Oil and the Underdeveloped Countries</u>
   (Boston: Beacon Press, 1969), pp. 320-321.
3. Tanzer, <u>op. cit.</u>, pp. 320-321.
4. Odell in Penrose, <u>op. cit.</u>, p. 291.
5. <u>The Times</u> (London), February 19, 1975; Odell in Penrose,
   <u>op. cit.</u>, pp. 291-292; Peter R. Odell, <u>Oil and World
   Power: A Geographical Interpretation</u> (New York: Tap-
   linger Publishing Company, 1971), pp. 86-87.
6. Odell in Penrose, <u>op. cit.</u>, p. 289.
7. Tanzer, <u>op. cit.</u>, p. 292.
8. <u>International Petroleum Encyclopedia 1971</u> (Tulsa,
   Oklahoma), pp. 146-153.
9. <u>Oil and Gas Journal</u>, June 26, 1972; <u>Oil and Gas Journal</u>,
   January 15, 1973. According to <u>Webster's New World Dic-
   tionary</u>, cretaceous refers to things "containing, composed of, or having the nature, of chalk;" it also refers
   to a geological period during which there was the development of chalk beds.
10. <u>Oil and Gas Journal</u>, January 28, 1974.
11. <u>The Wall Street Journal</u>, March 20, 1975.
12. <u>The Wall Street Journal</u>, June 16, 1975.
13. <u>World Oil</u>, August 15, 1977.
14. <u>Oil and Gas Journal</u>, January 6, 1975.
15. "Petroleum 2000" in <u>Oil and Gas Journal</u>, August 1977,
    p. 489.
16. <u>International Petroleum Encyclopedia 1977</u> (Tulsa, Okla-
    homa), pp. 129-134; <u>Petroleum Economist</u>, March 1977,
    p. 94; <u>World Oil</u>, August 15, 1977, pp. 64-65.
17. <u>Oil and Gas Journal</u>, September 19, 1977; <u>World Oil</u>,
    August 15, 1977, p. 64.
18. <u>World Oil</u>, August 15, 1977, pp. 64-65.
19. <u>Petroleum Intelligence Weekly</u>, July 4, 1977.
20. <u>The Financial Times</u> (London), April 6, 1978; <u>Oil and Gas
    Journal</u>, April 17, 1978; <u>Oil and Gas Journal</u>, May 1, 1978;
    <u>Petroleum Economist</u>, December 1977.
21. <u>Petroleum Intelligence Weekly</u>, February 21, 1977.
22. Central Intelligence Agency, <u>The International Energy
    Situation: Outlook to 1985</u> (Washington, D.C., April 1977),
    p. 11.
23. <u>Petroleum Economist</u>, March 1977, p. 94.
24. <u>Petroleum Intelligence Weekly</u>, August 29, 1977.
25. <u>Petroleum Economist</u>, July 1977, p. 279.
26. <u>Oil and Gas Journal</u>, September 19, 1977.

27. <u>Petroleum Intelligence Weekly</u>, October 17, 1977; <u>Petroleum Economist</u>, November 1977; <u>Petroleum Intelligence Weekly</u>, December 19, 1977; <u>Oil and Gas Journal</u>, December 19, 1977 and March 26, 1979.
28. <u>Oil and Gas Journal</u>, February 28, 1977 and December 26, 1977.
29. <u>Oil and Gas Journal</u>, February 28, 1977.
30. <u>Oil and Gas Journal</u>, March 14, 1977, p. 45; <u>Petroleum Economist</u>, March 1977, p. 94.
31. <u>Oil and Gas Journal</u>, April 17, 1978.
32. "Petroleum 2000," <u>op. cit.</u>, p. 489.
33. <u>Oil and Gas Journal</u>, February 28, 1977, pp. 85-92; <u>Petroleum Economist</u>, March 1977, pp. 94-95.
34. <u>Forbes</u>, July 1, 1977.
35. <u>The New York Times</u>, September 9, 1977.
36. <u>The Wall Street Journal</u>, October 17, 1977.
37. <u>Le Monde</u>, November 12, 1978.
38. <u>The Wall Street Journal</u>, December 8, 1977.
39. <u>The Journal of Commerce</u>, September 16, 1977.
40. <u>Petroleum Intelligence Weekly</u>, January 2, 1978.
41. <u>The Financial Times</u> (London), February 21, 1978; <u>The Wall Street Journal</u>, March 17, 1978.
42. David Gordon, "Mexico: A Survey," <u>The Economist</u>, April 22, 1978.
43. <u>The New York Times</u>, March 15, 1977.
44. <u>Ibid.</u>, March 21, 1977.
45. <u>Fortune</u>, April 10, 1978.
46. <u>The Wall Street Journal</u>, February 11, 1975.
47. <u>The Economist</u>, April 22, 1978; <u>Forbes</u>, July 1, 1977; <u>Fortune</u>, April 10, 1978.
48. <u>The Daily Star</u> (Beirut), October 18, 1974; <u>The Times</u> (London), February 19, 1975.
49. <u>The Daily Star</u> (Beirut), October 23, 1974.
50. <u>Ibid.</u>, May 16, 1975.
51. <u>Petroleum Intelligence Weekly</u>, September 27, 1976.
52. <u>Ibid.</u>, August 2, 1976.
53. <u>Forbes</u>, July 1, 1977.
54. Antonio J. Bermúdez, "The Mexican National Petroleum Industry," special issue of <u>Hispanic American Report</u> (San Jose, Calif.: Stanford University Press, 1963), pp. 246-247.
55. <u>The Christian Science Monitor</u>, January 14, 1975; <u>The New York Times</u>, October 15, 1974.
56. <u>The Christian Science Monitor</u>, February 17, 1977.
57. <u>Times of Americas</u>, January 21, 1976.
58. <u>The Wall Street Journal</u>, June 2, 1977, October 26, 1977 and December 5, 1977; <u>Petroleum Intelligence Weekly</u>, June 6, 1977, p. 5 and August 8, 1977, p. 9; <u>Petroleum Economist</u>, July 1977, p. 279 and February 1978, p. 70; <u>The Christian Science Monitor</u>, August 4, 1977; <u>The New York Times</u>, August 5, 1977, September 24, 1977, December 17, 1977, December 30, 1977, January 6, 1978 and May 4, 1978; <u>The Economist</u>, August 27, 1977, pp. 83-84; <u>The</u>

Journal of Commerce, October 19, 1977, October 25, 1977,
November 2, 1977, December 27, 1977, January 6, 1978 and
January 13, 1978; The Washington Post, November 25, 1977;
Oil and Gas Journal, December 19, 1977, p. 38 and April
17, 1978, p. 68; Petroleum Times, January 6, 1978, p. 10;
Business Week, January 9, 1978, p. 32.

59.   Petroleum Intelligence Weekly, October 17, 1977 and December 19, 1977; The Financial Times (London), April 6,
1978; Le Monde, April 12, 1978.

60.   The Journal of Commerce, November 30, 1977.

61.   Ibid., January 23, 1978; Financial Times (London), April
6, 1978; Oil and Gas Journal, September 19, 1977; Petroleum Economist, February 1978, p. 78.

62.   Financial Times, April 10, 1979; June 8, 1979; July 5,
1979; The Journal of Commerce, April 12, 1979; May 8,
1979; The Oil and Gas Journal, September 18, 1978; May 21,
1979; Petroleum Intelligence Weekly, December 18, 1978;
February 5, 1979.